Lecture Notes of the Institute for Computer Sciences, Social Informatics and Telecommunications Engineering 366

More information about this series at http://www.springer.com/series/8197

Eva Irene Brooks · Anthony Brooks ·
Cristina Sylla · Anders Kalsgaard Møller (Eds.)

Design, Learning, and Innovation

5th EAI International Conference, DLI 2020
Virtual Event, December 10–11, 2020
Proceedings

 Springer

Editors
Eva Irene Brooks
Aalborg University
Aalborg, Denmark

Cristina Sylla
University of Minho
Braga, Portugal

Anthony Brooks
Aalborg University
Aalborg, Denmark

Anders Kalsgaard Møller
Aalborg University
Aalborg, Denmark

ISSN 1867-8211   ISSN 1867-822X (electronic)
Lecture Notes of the Institute for Computer Sciences, Social Informatics
and Telecommunications Engineering
ISBN 978-3-030-78447-8   ISBN 978-3-030-78448-5 (eBook)
https://doi.org/10.1007/978-3-030-78448-5

This Springer imprint is published by the registered company Springer Nature Switzerland AG
The registered company address is: Gewerbestrasse 11, 6330 Cham, Switzerland

Preface

We are delighted to introduce the proceedings of the fifth edition of the European Alliance for Innovation (EAI) International Conference on Design, Learning & Innovation (DLI 2020). This conference brought together researchers, designers, educators, and practitioners from around the world to share their latest research findings, emerging technologies, and innovative methods in the areas of inclusive and playful designs and learning with digital technologies. The theme for DLI 2020 was "Sustainable development realizing human needs" targeting a conceptualisation of the effects and impact of digital technologies for, in a sustainable and playful way, fostering human beings in realizing their needs and ideas.

The technical program of DLI 2020 consisted of 14 full papers, which were presented across four main conference sessions. Session 1 – Digital Technologies and Learning; Session 2 – Designing for Innovation; Session 3 – Digital Games, Gamification and Robots; and Session 4 – Designs for Innovative Learning. The four tracks were chaired by Pedro Ribeiro, Vitor Carvalho, Eva Cerezo, and Maite Gil, respectively. Their contributions to the discussion related to the sessions were exceptional and created a productive online atmosphere with fruitful and constructive dialogues.

The paper presentations addressed new dimensions and key challenges and provided critical and innovative perspectives of employing digital technologies and games to develop and implement future design, learning, and innovation. This was reflected, among other things, by the paper which received the Best Paper Award of EAI DLI 2020, titled The Impact of a Digitally-Augmented Reading Instruction on Reading Motivation and Comprehension of Third Graders, by Pedro Ribeiro (Rhine-Waal University of Applied Sciences), Anna Michel (Rhine-Waal University of Applied Sciences), Cristina Sylla (University of Minho), Ido Iurgel (Rhein-Waal University of Applied Sciences), Wolfgang Müller (University of Education Weingarten), Christian Ressel (Rhine-Waal University of Applied Sciences), and Katharina Wennemaring (Rhine-Waal University of Applied Sciences) – congratulations on the award!

The collaboration with the Technical Program Committee co-chairs, Cristina Sylla and Anders Kalsgaard Møller, was essential for the successful planning and performance of the conference. We sincerely appreciated the coordination with Viltaré Platzner, the senior conference manager at the European Alliance for Innovation (EAI), and Imrich Chlamtac, the steering chair. We also appreciated the constructive cooperation with EAI ArtsIT 2020. We are genuinely thankful for the support of the Organizing Committee team: Anders Kalsgaard Møller and Cristina Sylla (Technical Program Committee co-chairs), João Martinho Moura (Publicity and Social Media chair), Jeanette Sjöberg (Workshops chair) and Camilla Finsterbach Kaup (Demos and Posters chair). We also acknowledge the outstanding work by the Technical Program Committee members. Last but not least, we are grateful to all the authors who submitted their papers to the DLI 2020 conference.

We strongly believe that the DLI conference provides a productive forum for researchers, designers, educators, and practitioners to discuss the cross-disciplinary field of digital technology and its implications on design, learning, and innovation. We also expect that future DLI conferences will provide a fruitful arena for knowledge exchange, as indicated by the contributions presented in this volume.

Eva Brooks

Organization

Steering Committee

Imrich Chlamtac   European Alliance of Innovation (EAI)
Eva Brooks   Aalborg University, Denmark

Organizing Committee

General Chair

Eva Brooks   Aalborg University, Denmark

Technical Program Committee Chair and Co-chair

Anders Kalsgaard Møller   Aalborg University, Denmark
Cristina Sylla   University of Minho, Portugal

Local Chairs

Tine Skjødt Andreasen   Aalborg University, Denmark
Jeanette Arboe   Aalborg University, Denmark

Workshops Chair

Jeanette Sjöberg   Halmstad University, Denmark

Publicity and Social Media Chair

João Martinho Moura   Polytechnic Institute of Cávado and Ave, Portugal

Publications Chair

Eva Brooks   Aalborg University, Denmark

Web Chair

João Martinho Moura   Polytechnic Institute of Cávado and Ave, Portugal

Demos and Posters Chair

Camilla Fausterbach Kaup   Aalborg University, Denmark

Technical Program Committee

Alejandro Catala	Universidade de Santiago de Compostela, Spain
Anders Kalsgaard Møller	Aalborg University, Denmark
André Rabe	Universidade do Vale do Itajaí, Brazil
Anthony Brooks	Aalborg University, Denmark
António Quintas-Mendes	Universidade Aberta, Portugal
Camilla Finsterbach Kaup	Aalborg University, Denmark
Camilla Gyldendahl Jensen	Aalborg University, Denmark
Chrystalla Neophytou	Open University, Cyprus
Clara Coutinho	University of Minho, Portugal
Cristina Sylla	University of Minho, Portugal
Digdem Sezen	Teesside University, UK
Dorina Gnaur	Aalborg University, Denmark
Elisabeth Lauridsen Lolle	Aalborg University, Denmark
Emma Edstrand	Halmstad University, Sweden
Eva Cerezo	University of Zaragoza, Spain
Evgenia Vassilakaki	Technological Educational Institute of Athens, Greece
Filomena Soares	University of Minho, Portugal
Günter Wallner	Eindhoven University of Technology, the Netherlands
Ido Iurgel	Rhine-Waal University of Applied Sciences, Germany
Jeanette Sjöberg	Halmstad University, Sweden
Lucia Amante	Open University, Portugal
Lykke Bertel Brogaard	Aalborg University, Denmark
M. Esther Del Moral	University of Oviedo, Spain
Maiga Chang	Athabasca University, Canada
Maitê Gil	University of Minho, Portugal
Marie Bengtsson	Halmstad University, Sweden
Mel Krokos	University of Portsmouth, England
Nikoleta Yiannoutsou	European Commission, Spain
Nuno Otero	Linnaeus University, Sweden
Pär-Ola Zander	Aalborg University, Denmark
Susanne Dau	University College of Northern Denmark, Aalborg, Denmark
Susanne Haake	University of Education Weingarten, Germany
Taciana Pontual Falcao	Universidade Federal Rural de Pernambuco, Brazil
Teresa Romão	Universidade Nova de Lisboa, Portugal
Thanasis Hadzilacos	Open University, Cyprus
Ute Massler	University of Education Weingarten, Germany
Vitor Sá	Universidade Católica, Portugal
Wolfgang Muller	University of Education Weingarten, Germany

Contents

Digital Technologies and Learning

The Impact of a Digitally-Augmented Reading Instruction on Reading Motivation and Comprehension of Third Graders

Pedro Ribeiro[1]([✉]), Anna Michel[1], Cristina Sylla[2], Ido Iurgel[1], Wolfgang Müller[3], Christian Ressel[1], and Katharina Wennemaring[1]

[1] Rhine-Waal University of Applied Sciences, Kamp-Lintfort, Germany
pr@hsrw.eu
[2] University of Minho, Braga, Portugal
[3] University of Education Weingarten, Weingarten, Germany

Abstract. A technical and conceptual framework is currently under development to augment the activity of reading at schools with digital media such as images, sounds and light effects. The technical framework is STREEN (Story Reading Environmental Enrichment). In this paper, we present IRIS, the pedagogical conceptual framework that describes how to employ STREEN in the classroom. We assume that STREEN/IRIS can motivate and foster reading comprehension of primary school students. We also describe an eight-weeks' study carried out with third-grade students using IRIS. This study follows a quasi-experimental pre-post design that took place at a German primary school with 56 students from three different third-grade classes to compare results between an IRIS instruction and two conventional reading instructions. The findings show that students in the experimental group improved in word and sentence comprehension and lowered their task error rate. Furthermore, their intrinsic reading motivation increased while extrinsic reading motivation decreased significantly.

Keywords: Augmented reading · Design-based research · Reader's theatre · Digital media · Children · Technology-based reading instruction

1 Introduction

1.1 Aim of the Research

Story Reading Environmental Enrichment (STREEN) is an augmented reading technical framework capable of supporting reading aloud activities at primary school by offering engaging features that promote vocabulary acquisition, reading fluency, reading comprehension and motivation. Here we present the Integrated Reading Instruction (IRIS) that uses STREEN to support students to become competent and motivated readers and teachers to use scaffolding strategies. IRIS is designed as a reader-centred instruction that attempts to meet the needs of every single student and to support students in

E. I. Brooks et al. (Eds.): DLI 2020, LNICST 366, pp. 3–25, 2021.
https://doi.org/10.1007/978-3-030-78448-5_1

achieving educational reading goals. It offers motivating tasks that promote the usage of multiple reading-related abilities. Ultimately, those tasks aim at activating the cognitive and motivational processes of reading that support students' comprehension and facing challenges to achieve competence in reading.

The investigation took place in a local primary school following a Design-Based Research approach that informed the design, implementation, and evaluation of the IRIS instruction and all the associated learning activities and technological tools.

The intervention was carried out for eight weeks, with 56 third-graders from three different classes to compare results between an IRIS instruction and two conventional reading instructions. The intervention aimed at: (i) assessing the impact of IRIS on the reading comprehension of 3rd graders, and (ii) assessing the impact of IRIS in students' reading motivation.

1.2 Background and Previous Work

In 2016 the international assessment of student's performance in reading literacy in the fourth grade [1] reported that only 43% of the students like to read. Since a positive attitude towards reading is crucial for the development of reading comprehension, it is essential to provide novel strategies to increase the effectiveness of the practices and to foster reading motivation. A reading instruction must provide explicit instruction in specific comprehension strategies and offer students a meaningful reading experience that promotes reading, interpretation and discussion of the text [2, 3]. Researchers and educational professionals are aware of the importance of digital technologies, and many would welcome new digital tools to support, complement and evolve reading instructional practices. As a response, the augmented reading technical framework STREEN (Story Reading Environmental Enrichment) is being developed to support and complement reading aloud activities at primary schools [4]. STREEN provides engaging functionalities to promote vocabulary acquisition, reading fluency and text comprehension.

The STREEN approach is related to similar attempts that use digital media to foster reading competences, including the intervention "Moved by Reading", where children read a story aloud and then move graphical elements, which correspond to the narrative [5], the project Gamelet that targets reading fluency in native and foreign languages by applying digital media-based gamification mechanisms [6], and the project LIT KIT that explores a cyber-physical system to transform read-aloud activities into a multimedia, mixed-reality experience [7]. STREEN differs from the referred approaches by stressing the story interpretation; it requires that children first collaboratively create digital media (e.g. images, sounds and light effects) that they will later employ when reading the story aloud, in front of others (see Fig. 1).

STREEN consists of a mobile application (the STREEN app) running on the student's tablet computers, a multimedia infrastructure comprising a projector, a surround sound system, smart bulbs, and a multimedia controller (the STREEN Controller) that runs on a computer server. We envision that such a media enriched environment will promote engaging and immersive reading experiences, thus increasing reading motivation and comprehension. Previous work on STREEN reported on the co-design process involving researchers, teachers, and children from a local primary school, in which the support of

Fig. 1. Primary school students interacting with STREEN in the classroom.

the acknowledged pedagogical activity Reader's Theatre with digital media was explored [4, 8].

In a preliminary study, we assessed students' acceptance of STREEN. The findings of the self-reported Unified Theory of Acceptance and Use of Technology (UTAUT) questionnaire [9] showed that most of the study participants were very interested in the technology and enjoyed reading with the STREEN app; they thought it to be helpful and showed the intention to use it in the future. The same findings were reported by the students that co-designed the STREEN prototype in participatory settings. Based on this evidence, the next step was to devise appropriate didactic methods and instructional practices to ensure sustainable motivation and reading comprehension.

2 The IRIS Instruction

IRIS stands for the Integrated Reading Instruction with STREEN and aims at foster-ing cognitive and motivational aspects of reading competence in upper primary school students. IRIS instruction was generated collaboratively with a third-grade teacher and researchers over six months, following design-based research methods [10]. The con-ceptual perspective of IRIS is that the development and growth of reading competence is a joint functioning of various cognitive, motivational, and social processes [11–13]. For an effective promotion of reading motivation and comprehension, research recommends providing multiple supportive elements [14–18]. For IRIS, we defined three supportive instructional elements: i) motivation-enhancing instructional practices, ii) motivating context, and iii) a balanced instructional structure.

2.1 Motivation-Enhancing Instructional Practices

The first element of the IRIS instruction is a set of motivation-enhancing instructional practices, designed to support students' psychological needs for relatedness, auton-omy, perceived competence, curiosity and involvement in reading, as well as to help them recognise the relevance of reading and orienting them towards mastery of reading competence [14, 17, 19–24].

Accordingly, in the IRIS classroom, the teacher provides students with motivating reading materials and tasks to entice their curiosity, involving them in reading activities and tasks that allow them to experience competence and control in reading and learning. The IRIS teacher promotes students working collaboratively on achieving mastery in reading. Besides, the teacher provides opportunities for success through the transparency of goals and by helping students to evaluate their success and failures appropriately.

Providing choice, giving motivational and constructive formative feedback and assessments, offering self-reflection and -evaluation are other practices determined to support perceived competence, relatedness and autonomy, as well as mastery orientation and reading relevance. IRIS also adds the practice of scaffolding, which involves teaching and letting students apply reading strategies to influence reading self-efficacy and reading comprehension directly. Supporting students' recognition of the relevance of reading is the last practice in IRIS.

2.2   Meaningful and Authentic Context

Providing meaningful and authentic context is a commonly recommended approach to engage and motivate students in learning as well as to support their achievements [25–27]. The second element in the IRIS instruction is the motivating context, which consists of three components: 1) Reader's Theatre, 2) mini-lessons and 3) STREEN.

Reader's Theatre

Reader's Theatre is a holistic approach for reading instruction, which provides teachers with a set of effective research-based pedagogical methods and reading activities to support students in developing reading fluency, vocabulary development, interest, and confidence in reading [28–36]. Built upon positive social interactions, Reader's Theatre is a motivating and engaging activity. It provides students with a variety of authentic reading tasks, which constitute a meaningful and authentic purpose for reading and gaining comprehension. In Reader's Theatre, students read a text in the form of a script. They choose roles and practice with others that have the same script, for the final Reader's Theatre performance. Reader's Theatre motivates students to read fluently and expressively, which is achieved through repeated reading in the regular reading sessions. The students are encouraged to use their voice and read expressively to convey the content of the text (messages, characters, plots, emotions). This additionally challenges the readers' imagination and helps the audience to understand the text and the story better. For expressive reading, students have to understand the text.

Mini-lessons

In order to improve reading comprehension and performance, students need specific tools [2, 13, 37–39]. Reading strategies taught in mini-lessons is a fundamental element provided in the IRIS classroom to experience pleasure, fun and curiosity, the feeling of competence and control, and to perceive the relevance of reading. In the process of planning and practising oral reading performance, the IRIS teacher conveys reading strategies to students in small teaching units that students can directly apply [40–42]. Teacher's support reduces, and the responsibility is gradually shifted from the teacher to the students as they become more competent and can apply the strategies independently [43].

STREEN

STREEN is the third contextual component offered in the IRIS classroom. As described above, STREEN consists of a digital-media reading environment, in which students collaboratively engage in constructing a shared understanding and knowledge of the reading materials in order to develop motivation and reading comprehension. STREEN provides a variety of technological features to support students accomplishing instructional tasks, applying strategies, reading digital texts, reflecting on reading, learning unknown vocabulary, expressing knowledge, documenting their work, and performing Reader's Theatre. It is also a tool for teachers to present instructional content, scaffold strategies, and evaluate students. In the following, we detail the main STREEN features:

Story Script: Allows the visualisation of the script and the touch-based text navigation. Additionally, in order to support children, it displays the line numbers and groups the script lines per role (see Fig. 2).

Highlight: Allows children to mark the text. The primary function of this feature is to draw attention to the lines children have to read aloud. There is a predefined highlight colour for each role. It can be alternatively used in the context of a word study strategy, such as finding unknown words (see Fig. 2).

Modelled Reading: STREEN provides a recorded modelled reading for each text. Once this functionality is activated, students can hear a human voice reading the story fluently. Children can select if they want to hear the complete story or only a part of the story. Additionally, by tapping a word, children can hear a text-to-speech generated audio and understand how the word sounds when read isolated (see Fig. 2).

Fig. 2. Story script with highlighted text (left); Story script with Modelled Reading enabled (right).

Word Cloud: Allows children to select words that will be clustered and displayed in a projection screen in the classroom. When children select a word, the word becomes surrounded by a red shape. Following the tag-cloud-like data visualisation [44], the Word Cloud provides an aggregate of word-selected statistics, displaying the word frequency information via font size (see Fig. 3).

Fig. 3. A student sending unknown words to the WordCloud.

Lexicon: An integrated lexicon provides a meaning for words that are likely to be unknown by 3rd-grade children; otherwise, STREEN guides children to obtain the meaning of the word on their own, suggesting, for example, to reread the sentence.

Authoring: Allows children to associate a word of the text with a produced digital media effect. By selecting a word, children access one of the pivotal features of STREEN, which support children with drag and drop mechanisms to compose an illustration by selecting, moving, scaling, mirroring and deleting 2D graphical representations of characters and other elements of the narrative from an integrated library. Furthermore, it also offers the possibility to author the ambient light and the soundscape. This authoring process aims to stimulate children's reading comprehension [5] it requires reading carefully and working with predefined digital media elements (DME) in order to produce an illustration and other DME that match the narrative. The produced digital media effect can also be discussed in different group settings (e.g. pairs, whole-class discussion) and can also serve as an outcome that can be evaluated by the teachers. Children can at any stage of the authoring process experience their work by projecting the illustrations in the projection screen and by controlling the light and sound system installed in the room (see Fig. 4).

Triggering: After finishing the authoring process, all the augmented words (with the created digital media) appear underlined in the script view. Afterwards, children can trigger enriched words, augmenting their reading with DME. There are two interaction options to trigger the effects; children can choose between tapping the word or sliding their finger below the augmented words (see Fig. 4).

Fig. 4. STREEN app authoring environment (left); Students using the STREEN app to read and trigger digital media effects previously authored (right).

Book Cover: Uses the Triggering feature to present a digital book cover (see Fig. 5). All the STREEN texts have a predefined book cover that supports the teacher in encouraging students to predict the story, a practice that positively impacts the student's comprehension of the text [2].

Record, Listen, Reflect (RLR): Allows children to record their reading performance, listen to their records, reflect on mistakes they have done and define improvement goals (see Fig. 5). Children are guided by questions, which allow them to assess their performance, (e.g. Did I pronounce every word correctly? Have I paid attention to punctuation and its meaning? Have I read at a reasonable pace?).

Fig. 5. Students make predictions of the story (left); a student using the functionality "Record, listen, reflect" to improve his reading fluency (right).

For the integration of STREEN into the reading instruction, we determined principles based on the current research [45–50]. Each student had a tablet set up with the STREEN app that allowed her/him to read digital Reader's Theatre scripts, study unknown words, practice reading fluency, collaborate with peers, create and trigger digital media effects, and discuss and present their performance and outcomes. The IRIS teacher scaffolds the use of the STREEN features one at a time according to students' needs and gradually hands over the responsibility to the students as they become acquainted with a particular feature. In practice, the teacher first explains and presents to students how to operate the tablet and certain STREEN features, and after that, the students receive tablets and try out the application. To avoid distractions and to enable a smooth use of STREEN and the creation of digital media effects, teachers and students set rules and regulations for how to behave, work, and communicate with each other when using STREEN. The students are allocated sufficient time to practice and use the STREEN app.

2.3 Balanced Instructional Structure

The third and last condition of the IRIS framework is the balanced instructional structure [26, 51, 52]. Following the baseline offered by the classical Reader's Theatre [35, 36], this framework provides a contextual structure that helps in the design of a weekly reading instruction, ideally divided through five days, with an hour per day. During the week, students get familiar with the theatre script, plan the performance, rehearse, and execute the final performance in front of an audience. At the day-level, the targeted instruction is divided into three phases: 1) getting started, 2) elaboration, and 3) fading out [52]. Each phase is subdivided into sequences (see Fig. 6). This granular way of structuring the day provides a guideline for the teacher to plan the IRIS instruction and easily integrate the multiple supports that IRIS proposes for fostering reading motivation and comprehension.

The Getting Started Phase
The IRIS instruction begins with catching students' attention by activating their prior knowledge and experience; asking for anticipation and predictions to trigger their curiosity, involvement, and to support perceived competence and collaboration. Afterwards, in order to support the value of reading and emphasise mastery goals, the teacher clearly communicates to the students the learning goals and their inherent challenges. In that way, the students learn what goals need to be achieved in order to become competent readers and which tasks they have to perform to achieve personal mastery goals.

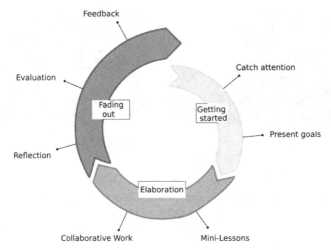

Fig. 6. The three phases of the instruction and the inherent activities.

With this knowledge, the students are prepared and can participate actively and without distractions towards achieving reading competence [53, 54].

The Elaboration Phase
The elaboration phase is sequenced in scaffolding strategies and offering collaborative work. Both sequences are part of the mini-lessons, in which the IRIS teacher conveys reading strategies and offers opportunities to apply them with STREEN and with other motivating teaching materials during the preparation for the Reader's Theatre performance.

The Fading Out Phase
The IRIS instruction ends with students reflecting on their learning process, progress and outcomes for an accurate self-evaluation. Feedback from the teacher, the peers and the teacher's assessment complete the final instructional phase. Ultimately, the teacher encourages students to participate in a voluntary homework assignment that consists of a small reading task related to the current instruction, which the students can accomplish for preparation for the next instruction and for maximising learning effects in strategy use and reading competence development. Over time, students internalise the instructional structure and can focus more on the content and participation rather than dealing with the question about what comes next, whether they will receive feedback or have time for reflection [52, 55, 56].

3   Implementation of IRIS at Primary School

As a first step for the integration of the IRIS framework, we established a partnership with a local primary school. Afterwards, a team composed of two researchers and one third grade teacher used the IRIS framework and worked for four months in the design of an eight-weeks IRIS instruction.

As a starting point, we identified and integrated cognitive components related to the reading competence, such as word recognition, reading fluency, and reading comprehension in the instructional framework elements. This identification and integration process was grounded in principles aimed at promoting reading comprehension and constrained by the specific requirements of each stage of Reader's Theatre. The following subsections present a model of a week instruction.

3.1 Monday

On the first day of the week, a new story is introduced. Research has shown that encouraging students to use their prior knowledge and to predict the content of a text can positively impact their comprehension of the text [2]. For that reason, on Mondays, the catch attention activity starts by activating relevant background knowledge about the story's topic, followed by an activity lead by the teacher, in which students are asked to anticipate and predict the plot of the narrative based on the analysis of the book cover that is displayed on the STREEN system (see Fig. 5). When a new text genre is introduced, e.g., fairy tales or fables, the teacher informs students about the essential characteristics of the text genre. After that, the teacher presents students the weekly schedule in which the week and daily goals are presented and facilitates a discussion aiming at the common understanding of the goals. In the elaboration phase, the teacher models reading aloud, and the students listen attentively. That can be done by having the teacher reading the script directly with the STREEN app or by allowing the students to hear a modeled reading through the STREEN app. Since it is crucial that students have the possibility to clarify unknown words at an early stage, STREEN provides a functionality, named WordCloud, that allows students to mark unknown words in the text and send them to a projection screen (see Fig. 3). The Word Cloud aims at supporting teachers and students in the activity of identifying, grouping and visualising unknown words, as well as to support the learning and application of strategies to discover the meaning of the words such as using an incorporated lexicon or context analysis. In the fading out phase, the students reflect on their reading process and progress and receive feedback from their peers and the teacher. At the end, students are encouraged to read the full script at home in order to achieve a deeper understanding of the narrative and to be better prepared to choose a role in the following day.

3.2 Tuesday

On Tuesday, the students choose their role and work on the common understanding of the text. The instruction starts by activating prior knowledge about the characters which are displayed on the projection screen and communicating the goals and challenges of the day. In the mini-lesson, the teacher forms small groups in which the students then choose and highlight their roles using the STREEN app (see Fig. 7). Following this, the students use the STREEN app to create a limited amount of digital media effects for their lines (see Fig. 7). This authoring process is targeted to support the learning and application of comprehension strategies such as Visualization, Questioning and Think-aloud [2]. While authoring, children have to use their imagination and develop their capacity for generating questions, which help them to understand the text they have to "illustrate" and complement with digital media effects (e.g. images, sounds and light effects). Additionally, children have to discuss and collaborate in order to guarantee the coherence among the different digital media effects created by the different students (see Fig. 7). This cooperative activity requires children to learn and use the think-aloud strategy. The unit ends with the reflection and evaluation phase. To practice fluency, the students are asked to read their role at home.

Fig. 7. Students marking their roles (left); students creating and discussing the DME (right).

3.3 Wednesday

The third day of the week is dedicated to finalising the authoring process. In the catch attention phase, the teacher uses the STREEN app to read and trigger digital media effects, which purposefully present errors and incoherencies. For instance, the teacher uses illustrations that do not match the text. This way, the teacher aims to stimulate children's awareness of digital media effects problems and to foster their capacities to analyse and constructively criticise digital media effects created by the other students. It also shows the teacher how well the students have understood the text, e.g. when they identify the mismatch between text and pictures. It follows the communication of the daily goals, which is deepening the text comprehension through the authoring process. In the elaboration phase children refine and finish the digital media effects. Before the reflection and feedback routine, the students use the STREEN app to read and to trigger digital media effects. By doing so, students practice reading in a group, the reading fluency and automaticity of triggering the effects in the right moment. In the last point of the fading out phase, the students are asked to read their role at home in order to improve their reading fluency and to increase their confidence.

3.4 Thursday

This is the penultimate day, the rehearsal day that precedes the final performance. Since the catch attention phase relates to the daily goals, the teacher reads a part of the script with a poor reading fluency performance and encourages their students to identify the problems, e.g., wrong prosody or tempo. Afterwards, the teacher communicates the goals and challenges of the day. In the elaboration phase, students are invited to use the STREEN app functionality "Record, listen, reflect." With this functionality, the students can listen to the pre-recorded reading aloud of the text either wholly or line by line.

Additionally, the "Record, listen, reflect" functionality allows them to record reading, listen to their records, identify/reflect on mistakes and define goals for improvement (see Fig. 5). With this approach, students are encouraged to engage in an improvement cycle targeted towards the mastery of oral reading fluency. The last action in the elaboration phase is the final rehearsal where students read their parts while triggering the digital media effects. The teacher can interrupt the performance to provide direct feedback and instructions. The unit finishes with reflection and feedback routine. This routine is used to give and receive feedback from the peers and the teacher. Once again, the students are invited to rehearse reading their role lines at home.

3.5 Friday

On Friday, the reading instruction is dedicated to the final performance in front of an audience previously selected by the students (see Fig. 4). In the catch attention phase, the teacher does warm-up exercises with the students to prepare them for reading aloud, to ease arousal and increase concentration. After the day's goal is presented, it is the students turn to perform their reading aloud, which they start after the technical check of the STREEN system. Once the students have performed the Reader's Theatre, they receive direct feedback from the audience and the teacher. The audiovisual recording of the final performance can be used to support reflection and feedback. The instruction ends with students' reflection and evaluation of the week.

4 Research Questions

This investigation aimed at answering the following research questions:

1. What is the impact of the use of IRIS on the reading motivation of 3rd graders?

    a. What is the impact of the use of IRIS on the intrinsic reading motivation of 3rd graders?
    b. What is the impact of the use of IRIS on the extrinsic reading motivation of 3rd graders?
    c. How did the levels of reading motivation of IRIS' students evolve when compared to the conventional reading instructions students?

2. What is the impact of the use of IRIS in student's reading comprehension?

    a. What is the impact of the use of IRIS in student's reading comprehension at the word level?
    b. What is the impact of the use of IRIS in student's reading comprehension at the sentence level?
    c. What is the impact of the use of IRIS in students' overall reading comprehension (word and sentence level)?
    d. Does the prior tablet experience affect the impact of IRIS on reading comprehension?

e. How did the levels of reading comprehension of IRIS' students evolve when compared to the conventional reading instructions students?
f. How did the levels of reading comprehension of IRIS' students evolve when compared to the national baseline?
g. How did the error rates levels of IRIS' students evolve when compared to the reading instructions students?

5 Research Study

To answer these research questions, we carried out an eight-weeks comparative study. We focused on third-grade students because it is crucial to start early to counteract reading comprehension and motivation problems, which increase in particular from the upper primary school grades [12, 57, 58]. At this stage, students are supposed to have acquired the fundamental reading skill of fluently decoding the text that makes them capable of evolving their reading comprehension and starting applying reading strategies. Teaching students reading comprehension is an important goal in upper primary school grades. The German curriculum foresees that students have acquired the basic reading skills by the beginning of the third year of school and can then focus on advanced skills (reading competence).

5.1 Participants

The study was carried out with 56 students from three third-grade classes from a primary school in Germany. The children were aged between 8 and 9 years old ($M = 8.32$, $SD = 0.47$); 49% were female. One class (20 children) was assigned to the experimental group using the IRIS instruction, another class (16 children) was assigned to the control group using the Reader's Theatre instruction (RTI), the third class (20 children) was assigned to the control group using a conventional reading instruction (CRI). Two children in the IRIS group and one child in the RTI group could not complete the demographics questionnaire, leaving 18 children in the IRIS group, 15 children in the RTI group and 20 in the CRI group. Details of demographics and tablet use information of the study participants are presented in Table 1.

5.2 Design

The study used a quasi-experimental design that included one experimental group (IRIS integrated reading instruction with STREEN) and two comparison groups (traditional Reader's Theatre instruction and conventional reading instruction).

5.3 Assessment Materials

Reading Comprehension
Reading comprehension was assessed in a pre and post-test with a paper-pencil normalised ELFE II test [59]. This test is divided into three subtests, word comprehension,

Table 1. Participant's demographics and computer literacy.

	Full sample	IRIS	RTI	CRI
Female	49.1%	50.0%	53.3%	55.0%
Age	8.32	8.44	8.20	8.30
Birthplace				
Germany	86.8%	94.4%	73.3%	90.0%
Other	13.2%	5.6%	26.7%	10.0%
Parents' birthplace				
Germany	46.2%	38.9%	33.3%	63.2%
Mixed	21.2%	16.7%	33.3%	15.8%
Other	32.7%	44.4%	33.3%	21.1%
Language at home				
German	49.1%	50.0%	26.7%	65.0%
Mixed	41.5%	38.9%	60.0%	30.0%
Other	9.4%	11.1%	13.3%	5.0%
Daily tablet use				
Never	34.6%	33.3%	53.3%	21.1%
Less than 30 min	26.9%	27.8%	13.3%	36.8%
30 to 60 min	9.6%	16.7%	0.0%	10.5%
1 to 2 h	17.3%	11.1%	20.0%	21.1%
More than 2 h	11.5%	11.1%	13.3%	10.5%

sentence comprehension and text comprehension. The word comprehension subtest comprises 75 items. In this subtest, each item presents one picture, and four words and students are required to mark the word that matches the picture. The sentence comprehension subtest consists of 36 items. The items present sentences in which a suitable word is missing and has to be selected and inserted from five alternatives offered. The text comprehension subtest includes 26 short text passages with associated multiple-choice questions. The task is to answer the question by selecting the appropriate one from the four answer alternatives offered. Students from the three study groups were given all three subtests. For the analysis, the text comprehension subtest was removed on account of its invalidity (ELFE II Manual). The data collected indicated that the processing of the subtest was invalid. The students of the experimental group were impeded completing the test by several internal circumstances.

Reading Motivation
Reading motivation was assessed through a survey designed to assess (i) intrinsic and extrinsic reading motivation (incentives within the reading process), (ii) reading self-efficacy (ability to perform reading well), and (iii) the importance of reading (perceived

significance and utility of reading). As theorised by Schiefele, the last two are the antecedents of reading motivation [60]. Intrinsic reading motivation refers to curiosity (interest related to the reading of particular text genre and topic) and involvement (experiences of enjoyment, deep absorption, and imagination, for example, getting involved with the characters in the story). The extrinsic reading motivation encompasses two dimensions: competition (outperform other students in reading) and recognition (get praise for good reading). Due to the lack of a uniform operationalisation of reading motivation and the inexistence of a standardised reading motivation questionnaire that covers all components of reading motivation, we combined the items from different standardised reading motivation questionnaires. Intrinsic and extrinsic reading motivation, as well as reading importance, were measured through the Reading Motivation Questionnaire for Elementary Students (RMQ-E) [61] and the Reading Motivation Questionnaire (RMQ) [62]. The Scale for Self-Concept in Reading [63] served to assess students' perception of reading self-efficacy. Our questionnaire initially contained 33 items and was applied in a pre-pre test to 16 students from another third-grade class of the same school who did not participate in the study to measure the questionnaire's reliability. The factor-analytical evaluation of the data showed that the questionnaire had a high reliability of cronbach's alpha .815 to measure the composition of reading motivation. A look at individual items showed that this could be increased to .839 by excluding some items. Since we also aim to measure the influence of IRIS on individual components of reading motivation, we have also performed the factor analysis for the respective reading motivation components. This has also given us a deeper insight into which items should be removed. The survey consisting of 29 items was conducted with the paper-pencil method. All items were worded in the standard statement format (e.g. "I read because reading is fun") requiring the participants to indicate how these statements applied to them on a four-point rating scale (1 = no, 2 = rather no than yes, 3 = rather yes than no, 4 = yes).

5.4 Procedure

From the beginning of the school year 2018/19 until mid-December 2018, we had several organisational meetings with the school director and the three teachers responsible for the experimental and the control groups to explain our research and to plan the study. Before Christmas holiday 2018, we distributed the informed consents describing the research procedures and data acquisition to parents. Parents and children were clearly informed (and agreed to) that the researchers might use and share data, including video, images, in publications or presentations for academic purposes. After parents signed and returned the consent forms and all children assented to participate in the study, we briefed the three teachers responsible for the experimental and the control groups about how to administer the data collection. Prior to the beginning of the intervention, the collection of the pre-test data was carried out within one week for the three groups, in three consecutive days, with the first-day directed for reading motivation, the second day for demographics and computer literacy and the third day for reading comprehension. In order to ensure anonymisation, all the questionnaires were coded by the teachers.

Following the pre-test week, the three groups engaged in a period of eight weeks of classroom instruction, in a total of 40 daily sessions, each with the duration of 60 min, carried out between 10:00 AM and 11:00 AM.

The teacher of the students in the experimental group was trained in IRIS instruction and assisted by one of the researchers responsible for providing support in setting up the STREEN infrastructure. Since the IRIS instruction follows a weekly cycle, a week in advance, the teacher received all the necessary materials to conduct the upcoming week of instruction. These materials always included a printed version of the goals for the week (see Fig. 6), to be posted in the classroom, the reading fluency checklist to support the Record, Listen, Reflect functionality, and for each student and teacher, a printed copy of the week plan and an updated version of the STREEN application comprising the story with related digital media running on a 10.1-in. tablet computer. Every week, reading strategies were gradually introduced to the students and applied throughout the week as tools for developing reading fluency and reading comprehension. The teacher was additionally provided with a printed procedural instruction, which included teachers and students actions, tasks for the students, and explanations on how to use specific features of STREEN. Finally, in the first week, we also provided a list of rules on how to use STREEN in A1 size format to be posted in the classroom.

For each of the eight-weeks, we selected one story, in the first four weeks, we used fairy tales, and in the following weeks fables. In order to address the gradual release principle, the STREEN features were introduced during the first four weeks. In the first week, we also had to introduce children to the procedure inherent to readers' theatre.

The instructions in the two control groups were conducted in parallel in two other classes of the same primary school. The first control group used traditional Reader's Theatre instruction and the second class used conventional instruction for reading competence promotion. All groups read the same story in the same week. A printed copy of the eight-story scripts given to each student and teacher from the experimental group was provided to the two control group teachers in the pre-test week. The script text of these printed versions was formatted according to the digital script used by the experimental group, e.g. font type/size and line spacing.

Directly after the last week of instruction, the collection of post-test data was carried out within one week in four consecutive days, with the first day dedicated to reading motivation, the second day for demographics and computer literacy, the third day for reading comprehension and the fourth day for instruction feedback (only administered to the experimental group).

6 Results

Before the intervention, the teachers conducted pre-tests assessing students' reading comprehension and motivation. The post-tests were carried out the week following the last instruction. Two children in the IRIS group and three children in the CRI group could not complete the reading comprehension questionnaire, remaining a total of 18 participant children in the IRIS group, 16 in the RTI group and 17 in the CRI group. A similar situation happened with the reading motivation questionnaire, leaving 18 children in the IRIS group, 13 in the RTI group and 19 in the CRI group.

6.1 The Impact of IRIS on Reading Comprehension

To investigate the impact of IRIS on the reading comprehension, we first run tests to analyse the data collected from the ELFE II subtests indicated by the number of correct answers. For measuring the impact of IRIS on word comprehension, we ran a t-test with results showing that the word comprehension ($M = 46.89$, $SD = 2.09$) increased significantly after the eight-weeks IRIS instruction ($M = 54.00$, $SD = 1.58$, $t(17) = -6.65$, $p^{**} < .01$). In the same way, students' sentence comprehension ($M = 45.61$, $SD = 9.79$) increased significantly after the intervention ($M = 50.67$, $SD = 2.18$, $t(17) = -6.89$, $p^{**} < .01$). Regarding the overall performance in reading comprehension (word and sentence comprehension), the IRIS class also showed significant improvement from the pre-test ($M = 46.00$, $SD = 2.20$) to the post-test ($M = 52.50$, $SD = 1.89$, $t(17) = -7.87$, $p^{**} < .01$). In terms of the impact of prior tablet experience on the overall performance in reading comprehension, the results indicate that the children with less table experience (less than 30 min per day) present a lower improvement ($n = 11$, $M = 48.00$, $M = 53.27$) when compared with children with more experience ($n = 7$, $M = 42.86$, $M = 51.29$); however, a two-way ANOVA with repeated measures indicated that this between-group difference was not statistically significant ($F(2) = 4.109$, $p = .06$).

6.2 The Impact of IRIS on Reading Comprehension in Comparison with Two Control Groups

As a starting point, we compare the competence of the three different classes before the treatment. The analyses of variance yielded no significant difference between the classes ($F(2) = 1.779$, $p^* > .05$). With regard to the overall reading comprehension, the classes improved during the intervention significantly ($F(2) = 287.505$, $p^{**} < .01$) and they differed significantly between each other ($F(2) = 6.337$, $p^{**} < .01$). The pairwise comparison reveals that the CRI ($M = 49.71$, $M = 60.53$) and RTI classes ($M = 43.19$, $M = 53.25$) differ significantly ($M = -6.899$, $p^* < .05$). Regarding the word comprehension, all groups scored significantly higher ($F(2) = 144.068$, $p^{**} < .01$). However, the differences between the groups were not significant ($F(2) = 2.861$, $p^* > .05$). For assessing the differences between the groups on the sentence comprehension, we ran the Kruskal-Wallis test because of the significance in the Levene's test. The three classes improved their levels of sentence comprehension after the intervention and the difference was significant between the groups ($H(2) = 8.453$, $p^* < .05$). The improvement of IRIS group was the lowest and differed significantly compared with CRI ($H(2) = 6.868$, $p^{**} < .01$) and RTI ($H(2) = 5.474$, $p^* < .05$).

In terms of the amount of items answered, the results revealed that the IRIS class ($M = 65.11$, $M = 75.28$) scored higher after the intervention. However, there was an even stronger increase in the CRI ($M = 72.12$, $M = 98.47$) and RTI classes ($M = 58.56$, $M = 85.00$). A two-way ANOVA with repeated measures indicated that the three groups scored significantly higher after the intervention ($F(2) = 294.178$, $p^{**} < .01$) and that groups differed significantly from one another ($F(2) = 20.132$, $p^{**} < .01$). The pairwise comparison shows that the CRI class significantly differs with RTI ($M = 13.513$, $p^{**} < .05$) and IRIS ($M = 15.100$, $p^{**} < .05$). Regarding the amount of errors, the results show that the IRIS class reduced the average amount of errors ($M = 2.78$,

$M = 1.44$), whereas the CRI class ($M = 2.29, M = 3.59$) and the RTI class ($M = 1.88,$ $M = 2.06$) increased the average amount of errors. In the last step of the analysis, we assessed the error rate that is calculated by dividing the number of error occurrences by the total number of items answered. The results revealed that the IRIS was the class that registered the strongest decrease in the error rate ($M = 0.051, M = 0.022$), followed by the RTI class ($M = 0.033, M = 0.026$). Contrary to the two previous groups, the CRI class slightly increased the error rate ($M = 0.033, M = 0.039$). The distribution of the difference between post error rate and pre error rate between the groups was significant ($H(2) = 7.248, p^* < .05$). The post-hoc tests reveal that the error rate difference in the IRIS class ($M = -.0288$) significantly differed from the CRI class ($M = .0056, H(2) = 2.681, p^* < .05$).

Finally, the normed data for the reading comprehension test shows that although the improvement of the IRIS class was the lowest, the students improved compared to the official reading comprehension baseline in Germany which is provided by a t-test ($t(17) = -2.78, p = .013$). In the pre-test, the experimental group scored 46% higher than the sample used to create the official baseline. In the post-test, the results from the experimental group were 52% higher than the normed sample.

6.3 The Impact of IRIS on Reading Motivation

To assess the impact of IRIS on reading motivation, we applied the non-parametric Wilcoxon test for dependent samples due to the lack of normal distribution of the data. For the composite reading motivation, the test reveals that reading motivation in IRIS students ($n = 18$) differ significantly after participating in the IRIS instruction ($M = 3.03, M = 3.15$), $z = -2.526, p^* < .05$. The intrinsic reading motivation did not improve significantly ($M = 3.53; M = 3.38$), $z = -1.438, p^* > .05$. On the dimension curiosity, students in the IRIS class scored lower after the intervention ($M = 3.47; M = 3.30$), $z = -1.658, p^* > .05$, so also was the impact on the dimension involvement ($M = 3.59; M = 3.49$), $z = -1.089, p^* > 05$. In contrast, extrinsic motivation decreased significantly ($M = 3.18, M = 2.28$), $z = -3.668, p^{**} < .01$. On the dimension competition ($M = 3.21; M = 2.26$), $z = -3.632, p^{**} < .01$ and recognition ($M = 3.06, M = 2.33$), $z = -2.289, p^* < .05$ students scored significantly lower after the intervention. Regarding reading self-efficacy, after the intervention there was a non-significant increase ($M = 2.89, M = 2.94$), $z = -.701, p^* > .05$ in students' perception of being able to read well. Students perceived the importance of reading slightly lower after the intervention, but not significantly ($M = 3.68, M = 3.57$), $z = -1.897, p^* < .05$.

6.4 The Impact of IRIS on Reading Motivation in Comparison with Two Control Groups

For comparing the changes in reading motivation between the classes, we ran the Kruskal-Wallis test for independent samples due to the lack of normal distribution in most of the values. For the composite reading motivation, we reverse coded the items competition and recognition. The Kruskal-Wallis test revealed that the changes between the groups differ significantly ($H(2) = 10.828, p^{**} < .01$). The composite reading motivation increased in IRIS ($M = 3.03, M = 3.15$) and RTI class ($M = 2.90, M = 2.95$); in contrast, decreased in CRI class ($M = 3.32$, $M = 3.19$). The post-hoc tests reveal that the difference is significant between CRI and IRIS ($H(2) = -3.181$, $p^{**} < .01$).

The intrinsic reading motivation, composed of dimensions curiosity and involvement changed after the intervention in all classes significantly ($H(2) = 6.328$, $p^* < .05$). Only the students from the RTI class improved slightly ($M = 3.10$, $M = 3.12$). The CRI class ($M = 3.64$, $M = 3.34$) differed significantly from the RTI class ($H(2) = -2.416$, $p^* < .05$). Looking at the individual dimensions, with regard to curiosity, it diminished in the students of all three classes and the changes did not differ significantly between the classes ($H(2) = 1.625$, $p^* > .05$). For involvement, the difference between the classes was significant ($H(2) = 11.841$, $p^{**} < .05$). Involvement in students from RTI class increased ($M = 3.05$, $M = 3.15$). In the other two classes this dimension diminished, particularly in the CRI class ($M = 3.70$, $M = 3.20$). The pairwise comparison shows that the changes in the CRI class significantly differ from the IRIS class ($H(2) = -2.750$, $p^* < .05$) and RTI ($H(2) = -3.076$, $p^{**} < .01$).

The extrinsic reading motivation, composed of the dimensions competition and recognition decreased in all classes significantly ($H(2) = 7.749$, $p^* < .05$). The post-hoc tests show that the IRIS class ($M = 3.18$, $M = 2.28$) significantly differ from the RTI class ($M = 2.69$, $M = 2.48$), $H(2) = 2.554$, $p^* < .05$. Regarding competition, all classes decreased and the changes between the classes were significant ($H(2) = 12.620$, $p^* < .05$). The difference between the IRIS ($M = 3.21$, $M = 2.26$) and the CRI class ($M = 1.36$, $M = 1.16$) was significant ($H(2) = 3.501$, $p^{**} < .01$); whereas not between the IRIS and the RTI class ($M = 2.71$, $M = 2.35$, $H(2) = 2.158$, $p^* > .05$) and between RTI and CRI ($H(2) = 1.017$, $p^* > .05$). On the recognition dimension, the changes between the classes differ significantly ($H(2) = 10.024$, $p^* < .05$). The CRI class decreased ($M = 2.41$, $M = 1.12$) and differs significantly ($H(2) = -3.120$, $p^{**} < .01$) from the RTI class that increased on this dimension ($M = 2.62$, $M = 3.00$). There was no difference between CRI and IRIS ($M = 3.06$, $M = 2.33$), $H(2) = -.996$, $p^* > .05$). The difference between IRIS and RTI was not significant ($H(2) = 2.232$, $p^* > .05$).

For reading self-efficacy, the classes did not differ significantly ($H(2) = 1.000$, $p^* > .05$). Looking at the means reveals that there is a drop in reading self-efficacy in the CRI group ($M = 2.81$, $M = 2.79$). IRIS ($M = 2.89$, $M = 2.94$) and RTI ($M = 2.91$, $M = 2.93$) increased slightly. Reading importance in IRIS students ($M = 3.69$, $M = 3.57$) and CRI students ($M = 3.61$, $M = 3.04$) decreased. Only the students in the RTI class improved on this reading motivation component ($M = 3.10$, $M = 3.18$). A Kruskal-Wallis test revealed that the changes between the classes are significant ($H(2) = 6.618$, $p^* < .05$). The CRI class differed significantly from the IRIS ($H(2) = -2.215$, $p^* < .05$) and the RTI ($H(2) = -2.157$, $p^* < .05$). No significant difference was found between IRIS and RTI ($H(2) = .131$, $p^* > 0.05$).

7 Discussion

The results presented above reveal that the students that experienced the eight- weeks IRIS instruction showed gains in reading motivation and comprehension. Although the IRIS group was only able to outrange the control groups in some of the components (e.g. reading comprehension error rate, extrinsic reading motivation), the results are promising. The results indicate that advancing the reading instruction with the STREEN digital-media technology, integrating sound practices for promoting motivation and text comprehension, can support the development of reading comprehension and motivation in third-grade students.

An important aspect is that it requires time to learn how to use the technology properly in an educational context [23, 48]. In fact, in this study, we observed that IRIS students less experienced with tablet devices improved poorly in reading comprehension when compared with more experienced children. This does not forcefully mean that we need an explicit training period before starting the instruction, but certainly implies changes in the planning of the gradual release model. It will also imply that future studies integrate an additional measurement point in time directly after the familiarisation period in which teachers and students get used to STREEN. Additionally, even though we acknowledge the potential of the technology to create rich environments that can motivate students and enhance learning, we are also aware of its harmful potential to distract students and negatively influence learning [23, 45]. Furthermore, very often, the gains in motivation are due to the novelty of the technological tools, and there is a tendency of decrease in motivation with time [23, 45, 50]. Our investigation aims at better understanding the influence of time on motivation. It may also be that the high motivation to use the technology, in the beginning, has slowed down the development of reading comprehension. Accordingly, future study designs need to consider the integration of more than two measurement points.

It is also valid to say that time is needed to get familiarised with all the novelties offered by the IRIS concept. The students had to adapt to the new routine, and so did the teacher. For instance, the implementation of Reader's Theatre was certainly an additional challenge at the beginning of the eight-weeks instruction. Over time, the students became much more confident in using the technology and in mastering different aspects of the IRIS instruction.

Finally, the teachers' influence may have created a bias. The influence of teachers on students has been known for a long time. First, teachers carried out the tests, and this might have affected the results. For example, it was surprising that the IRIS class scored worse in the text reading comprehension post-test than in the pre-test. A noticeable number of students had not got very far in the test, which suggests that something had happened in the last subtest of reading comprehension. Second, the students and the teacher might as well felt too much pressure to show good results, negatively impacting the test results. Finally, we need to consider the influence of teachers on students' learning and gaining competencies, motivation, emotions. For that, we will need to evaluate IRIS with different teachers and collect data from students about their attitude towards their teachers and consider it in the analyses. The results from the comparison of the IRIS class with the normed data from the ELFE-test strengthen this assumption.

8 Conclusion

In this paper we have presented the instructional framework IRIS, a framework that was created based on the existing research literature and that was developed through a close cooperation between researchers and a third-grade teacher from a German primary school. IRIS provides a holistic approach for fostering the reading competence of upper primary school students. This approach allows children to become competent and motivated readers. IRIS uses the potential of digital media to promote reading motivation and comprehension through the systematic integration of STREEN into instructional practices providing a pedagogical-media concept for digital media-based teaching and development of reading competence. Additionally, we have presented an exemplary IRIS instruction that was created for third-grade students. We have also described an eight-weeks comparative study aimed at understanding the impact and quality of the IRIS instruction framework and reported results, which indicate positive effects on students' reading comprehension and motivation. These results show some possible influences on the instruction and require further investigations. In future work, we plan to deepen the data analysis regarding the students and teachers' feedback and to use the insights from this study to redesign and enhance the IRIS framework and the STREEN technology.

Acknowledgements. We would like to thank the teachers and all the children that willingly work with us in this project. Pedro Ribeiro has been financed by the Rhine-Waal University of Applied Science, with a PhD grant.

References

1. Mullis, I.V.S., Martin, M.O., Foy, P., Hooper, M.: PIRLS 2016 International Results in Reading. http://timssandpirls.bc.edu/pirls2016/international-results/ (2017)
2. Duke, N.K., Pearson, P.D.: Effective practices for developing reading comprehension. J. Educ. **189**, 107–122 (2009). https://doi.org/10.1177/0022057409189001-208
3. Viana, F.L., Cadime, I., Santos, S., Brandão, S., Ribeiro, I.: The explicit teaching of reading comprehension. impact analysis of an intervention program. Revista Brasileira de Educação. 22 (2017). https://doi.org/10.1590/s1413-24782017227172
4. Ribeiro, P., Michel, A., Iurgel, I., Ressel, C., Sylla, C., Müller, W.: Empowering children to author digital media effects for reader's theatre. In: Proceedings of the 17th ACM Conference on Interaction Design and Children, pp. 569–574. ACM, New York (2018). https://doi.org/10.1145/3202185.3210793
5. Glenberg, A.M., Goldberg, A.B., Zhu, X.: Improving early reading comprehension using embodied CAI. Instr Sci. **39**, 27–39 (2011). https://doi.org/10.1007/s11251-009-9096-7
6. Sezen, D., Massler, U., Ribeiro, P., Haake, S., Iurgel, I., Parente, A.: Reading to level up: gamifying reading fluency. In: Sylla, C., Iurgel, I. (eds.) TIE 2019. LNICSSITE, vol. 307, pp. 3–12. Springer, Cham (2020). https://doi.org/10.1007/978-3-030-40180-1_1
7. Schafer, G., Fullerton, S., Walker, I., Vijaykumar, A., Green, K.: Words Become Worlds: The LIT ROOM, a Literacy Support Tool at Room-Scale. Presented at the June 8 (2018). https://doi.org/10.1145/3196709.3196728
8. Ribeiro, P., Michel, A., Iurgel, I., Ressel, C., Sylla, C., Müller, W.: Designing a smart reading environment with and for children. In: Proceedings of the Twelfth International Conference on Tangible, Embedded, and Embodied Interaction, pp. 88–93. ACM, New York (2018). https://doi.org/10.1145/3173225.3173274
9. Venkatesh, V., Morris, M.G., Davis, G.B., Davis, F.D.: User acceptance of information technology: toward a unified view. MIS Q. Manage. Inf. Syst. **27**, 425–478 (2003)
10. Anderson, T., Shattuck, J.: Design-based research a decade of progress in education research? Educ. Res. **41**, 16–25 (2012). https://doi.org/10.3102/0013189X11428813

11. Guthrie, J.T., Bennett, L., McGough, K.: Concept-Oriented Reading Instruction: An Integrated Curriculum to Develop Motivations and Strategies for Reading (1994)
12. McElvany, N., Kortenbruck, M., Becker, M.: Lesekompetenz und Lesemotivation. Zeitschrift für Pädagogische Psychologie **22**, 207–219 (2008). https://doi.org/10.1024/1010-0652.22.34.207
13. Rosebrock, C., Nix, D.: Grundlagen der Lesedidaktik und der systematischen schulischen Leseförderung. Schneider Verlag Hohengehren, Baltmannsweiler (2014)
14. Ames, C.: Achievement goals and the classroom motivation climate. In: Schunk, D.H., Meece, J.L. (eds.) Student Perceptions in the Classroom, pp. 327–348. Routledge (1992)
15. Guthrie, J.T., Mason-Singh, A., Coddington, C.S.: Instructional effects of concept-oriented reading instruction on motivation for reading information text in middle school. In: Guthrie, J.T., Wigfield, A., Klauda, S.L. (eds.) Adolescents' Engagement in Academic Literacy, pp. 155–215 (2012)
16. Guthrie, J.T., McCann, A.D.: Characteristics of classrooms that promote motivation and strategies for learning. In: Guthrie, J.T., Wigfield, A. (eds.) Reading Engagement: Motivating Readers through Integrated Instruction, pp. 128–148. Order Department, International Reading Association, 800 Barksdale Road, Newark, DE 19174-8159 Book No (1997)
17. Stipek, D.: Good instruction is motivating. In: Wigfield, A., Eccles, J.S. (eds.) Development of Achievement Motivation, pp. 309–332. Academic Press, San Diego (2002). https://doi.org/10.1016/B978-012750053-9/50014-0
18. Turner, J.C.: starting right: strategies for engaging young literacy learners. In: Guthrie, J.T., Wigfield, A. (eds.) Reading Engagement: Motivating Readers through Integrated Instruction, pp. 183–204. Order Department, International Reading Association, 800 Barksdale Road, Newark, DE 19174-8159 Book No (1997)
19. Assor, A., Kaplan, H., Roth, G.: Choice is good, but relevance is excellent: autonomy-enhancing and suppressing teacher behaviours predicting students' engagement in schoolwork. Br. J. Educ. Psychol. **72**, 261–278 (2002). https://doi.org/10.1348/000709902158883
20. Guthrie, J.T., Taboada, A.: Fostering the cognitive strategies of reading comprehension. In: Guthrie, J.T., Wigfield, A., Perencevich, K.C., Perencevich, K.C. (eds.) Motivating Reading Comprehension: Concept-Oriented Reading Instruction, pp. 87–112. Routledge (2004)
21. Guthrie, J.T., et al.: Growth of literacy engagement: changes in motivations and strategies during concept-oriented reading instruction. Read. Res. Q. **31**, 306–332 (1996). https://doi.org/10.1598/RRQ.31.3.5
22. Guthrie, J.T., Wigfield, A., Humenick, N.M., Perencevich, K.C., Taboada, A., Barbosa, P.: Influences of stimulating tasks on reading motivation and comprehension. J. Educ. Res. **99**, 232–246 (2006). https://doi.org/10.3200/JOER.99.4.232-246
23. Ryan, R.M., Deci, E.L.: Self-determination theory and the facilitation of intrinsic motivation, social development, and well-being. Am. Psychol. **55**, 68 (2000)
24. Streblow, L., Holodynski, M., Schiefele, U.: Entwicklung eines Lesekompetenz- und Lesemotivationstrainings für die siebte Klassenstufe. Bericht über zwei Evaluationsstudien. Psychologie in Erziehung und Unterricht **54**, 287–297 (2007)
25. Brown, J.S., Collins, A., Duguid, P.: Situated cognition and the culture of learning. Educ. Res. **18**, 32–42 (1989). https://doi.org/10.3102/0013189X018001032
26. Collins, A., Brown, J.S., Newman, S.E.: Cognitive Apprenticeship: Teaching the Craft of Reading, Writing, and Mathematics. Bolt, Beranek and Newman, Inc., Cambridge, Mass.; Illinois Univ., Urbana. Center for the Study of Reading (1987)
27. Herrington, J., Reeves, T., Oliver, R.: Immersive learning technologies: realism and online authentic learning. J. Comput. High. Educ. **19**, 80–99 (2007). https://doi.org/10.1007/BF03033421
28. Griffith, L.W., Rasinski, T.V.: A focus on fluency: how one teacher incorporated fluency with her reading curriculum. Read. Teach. **58**, 126–137 (2004). https://doi.org/10.1598/RT.58.2.1

29. Martinez, M., Roser, N.L., Strecker, S.: "I never thought I could be a star": a readers theatre ticket to fluency. Read. Teach. **52**, 326–334 (1998)
30. Moran, K.J.K.: Nurturing emergent readers through readers theater. Early Childhood Educ. J. **33**, 317–323 (2006). https://doi.org/10.1007/s10643-006-0089-8
31. Mraz, M., Nichols, W., Caldwell, S., Beisley, R., Sargent, S., Rupley, W.: Improving oral reading fluency through readers theatre. Read. Horizons **52**, 163–180 (2013)
32. Rasinski, T., Stokes, F., Young, C.: The role of the teacher in reader's theater instruction. Texas J. Lit. Educ. **5**, 168–174 (2017)
33. Rinehart, S.D.: "Don't think for a minute that i'm getting up there": opportunities for readers' theater in a tutorial for children with reading problems **20**, 71–89 (1999). https://doi.org/10.1080/027027199278510
34. Worthy, J., Prater, K.: "I thought about it all night": readers theatre for reading fluency and motivation. Reading Teacher. **56**, 294–297 (2002)
35. Young, C., Rasinski, T.: Implementing readers theatre as an approach to classroom fluency instruction. Read. Teach. **63**, 4–13 (2009). https://doi.org/10.1598/RT.63.1.1
36. Young, C., Stokes, F., Rasinski, T.: Readers theatre plus comprehension and word study. Read. Teach. **71**, 351–355 (2017). https://doi.org/10.1002/trtr.1629
37. Palinscar, A., Brown, A.L.: Reciprocal teaching of comprehension-fostering and comprehension-monitoring activities. Cogn. Instr. **1**, 117–175 (1984). https://doi.org/10.1207/s1532690xci0102_1
38. Pressley, M., et al.: Transactional instruction of comprehension strategies: the Montgomery County, Maryland. Sail Program. Read. Writ. Q. **10**, 5–19 (1994). https://doi.org/10.1080/1057356940100102
39. Snow, C.E.: Reading for Understanding: Toward an R&D Program in Reading Comprehension. RAND Corporation, Santa Monica (2002)
40. Hoyt, L.: Snapshots: Literacy Minilessons up Close. Heinemann, Portsmouth (2000)
41. Mathes, P.G., Babyak, A.E.: The effects of peer-assisted literacy strategies for first-grade readers with and without additional mini-skills lessons. Learn. Disabil. Res. Pract. **16**, 28–44 (2001). https://doi.org/10.1111/0938-8982.00004
42. Outsen, N., Yulga, St.: Teaching Comprehension Strategies All Readers Need. Scholastic Professioanl Books, New York (2002)
43. Pearson, P.D., Gallagher, M.C.: The instruction of reading comprehension. Contemp. Educ. Psychol. **8**, 317–344 (1983). https://doi.org/10.1016/0361-476X(83)90019-X
44. Kaser, O., Lemire, D.: Tag-cloud drawing: algorithms for cloud visualization. In: Proceedings of the Tagging and Metadata for Social Information Organization Workshop, 16th International World Wide Web Conference (WWW 2007), IW3C2, Banff, Canada (2007)
45. Aufenanger, St.: Zum Stand der Forschung zum Tableteinsatz in Schule und Unterricht aus nationaler und internationaler Sicht. In: Bastian, J., Aufenanger, St. (eds.) Tablets in Schule und Unterricht: Forschungsmethoden und -perspektiven zum Einsatz digitaler Medien, pp. 119–138. Springer Fachmedien Wiesbaden, Wiesbaden (2017). https://doi.org/10.1007/978-3-658-13809-7_6
46. Eickelmann, B.: Digitale Medien in Schule und Unterricht erfolgreich implementieren: eine empirische Analyse aus Sicht der Schulentwicklungsforschung. Waxmann Verlag, Münster (2010)
47. Lim, C.P.: Effective integration of ICT in Singapore schools: pedagogical and policy implications. Educ. Tech. Res. Dev. **55**, 83–116 (2007). https://doi.org/10.1007/s11423-006-9025-2
48. Mitzlaff, H.: Internationales Handbuch Computer (ICT), Grundschule, Kindergarten und neue Lernkultur. Schneider-Verlag Hohengehren (2007)

49. Senkbeil, M., Wittwer, J.: Die Computervertrautheit von Jugendlichen und Wirkungen der Computernutzung auf den fachlichen Kompetenzerwerb. In: Prenzel, M., Artelt, C., Baumert, J., Blum, W., Hammann, M., Klieme, E., Pekrun, R. (eds.) PISA 2006: die Ergebnisse der dritten internationalen Vergleichsstudie, pp. 277–308. Waxmann, Münster, New York, München, Berlin (2007)

50. Zwingenberger, A.: Wirksamkeit multimedialer Lernmaterialien. Kritische Bestandaufnahme und Metaanalyse empirischer Evaluationsstudien. Waxmann Verlag, Münster u.a. (2009)

51. Gagné, R.M., Briggs, L.J., Wager, W.W.: Principles of Instructional Design. Harcourt Brace College, Fort Worth (1992)

52. Meyer, H.: Unterrichtsentwicklung. Cornelsen (2015)

53. Jang, H., Reeve, J., Deci, E.L.: Engaging students in learning activities: it is not autonomy support or structure but autonomy support and structure. J. Educ. Psychol. **102**, 588–600 (2010). https://doi.org/10.1037/a0019682

54. Wang, M.-T., Eccles, J.S.: School context, achievement motivation, and academic engagement: a longitudinal study of school engagement using a multidimensional perspective. Learn. Instr. **28**, 12–23 (2013). https://doi.org/10.1016/j.learninstruc.2013.04.002

55. Doyle, W.: Ecological Approaches to Classroom Management. In: Evertson, C.M., Weinstein, C.S. (eds.) Handbook of Classroom Management: Research, Practice, and Contemporary Issues, pp. 97–125. Routledge, New York (2011)

56. Stipek, D., Chiatovich, T.: The effect of instructional quality on low- and high-performing students. Psychol. Sch. **54**, 773–791 (2017). https://doi.org/10.1002/pits.22034

57. Guthrie, J.T., Hoa, A.L.W., Wigfield, A., Tonks, St.M., Humenick, N.M., Littles, E.: Reading motivation and reading comprehension growth in the later elementary years. Contemp. Educ. Psychol. **32**, 282–313 (2007). https://doi.org/10.1016/j.cedpsych.2006.05.004.

58. Philipp, M.: Lesesozialisation in Kindheit und Jugend: Lesemotivation, Leseverhalten und Lesekompetenz in Familie. Schule und Peer-Beziehungen. Kohlhammer, Stuttgart (2011)

59. Lenhard, W., Lenhard, A., Schneider, W.: ELFE II. Ein Leseverständnistest für Erst- bis Siebtklässler - Version II. Manual. 1. Auflage. Hogrefe, Göttingen (2017)

60. Schiefele, U., Schaffner, E., Möller, J., Wigfield, A.: Dimensions of reading motivation and their relation to reading behavior and competence. Read Res Q. **47**, 427–463 (2012). https://doi.org/10.1002/RRQ.030

61. Stutz, F., Schaffner, E., Schiefele, U.: Measurement invariance and validity of a brief questionnaire on reading motivation in elementary students. J. Res. Reading **40**, 439–461 (2017). https://doi.org/10.1111/1467-9817.12085

62. Schiefele, U., Schaffner, E.: Factorial and construct validity of a new instrument for the assessment of reading motivation. Read. Res. Q. **51**, 221–237 (2016). https://doi.org/10.1002/rrq.134

63. Faber, G.: Ganztagsangebote im projekt „Schule im stadtteil "der stadt hannover. Eine empirische Bestandsaufnahme sowie Analysen zu ausgewählten Schüler-und Kontextmerkmalen in dritten und vierten Grundschulklassen. Leibniz Universität Hannover, Philosophische Fakultät: Institut für Pädagogische Psychologie (2010)

Explore the Effects of Usefulness and Ease of Use in Digital Game-Based Learning on Students' Learning Motivation, Attitude, and Satisfaction

Chun-Hsiung Huang[⊠]

Department of Digital Content Design, Ling Tung University, Taichung, Taiwan
huangch@teamail.ltu.edu.tw

Abstract. This study actually produced and developed a computer role-playing game which aimed at 108 first-year college students. It discovered the relationship between the perceptual usefulness and ease of use of digital game-based learning and students' learning motivation, learning attitude, and learning satisfaction. In this study, a questionnaire survey method was applied to analyze the relationship between the research variables and the input hypothesis verification for the 108 questionnaire data recovered. The following conclusions were obtained: (1) "Usefulness" positively affects "learning motivation," "learning attitude," and "learning satisfaction." (2) "Ease of use" does not affect "learning motivation," "learning attitude," and "learning satisfaction." (3) The research infers that students are now familiar with the operation of computer role-playing games. Whether it is easy to use is not an important consideration. However, in terms of teaching strategies, teachers should pay attention to the usefulness of learners' perceptions. If the perceptions are useful, teachers can greatly increase the chances of success in teaching, and can enable students to have equivalent learning motivation, attitude, and satisfaction.

Keywords: Digital Game-Based Learning (DGBL) · Learning motivation · Learning attitude · Learning satisfaction

1 Introduction

In terms of the concept of situated teaching, the meaning of knowledge cannot be isolated from the context. The computer role-playing games applies a variety of level designs to form a rich and reasonable storyline. Through the exploration, users can look for information and try to complete the tasks given in the game. The computer role-playing game makes the players deeply feel that they are the characters in the stories and survive in the virtual games as the characters. What attracts players is the process of story development and the curve of character development, as well as the relationships they build up with other characters in the game world.

The Digital Game-Based Learning (DGBL) system can provide learners with an active learning environment and opportunity. It allows learners to immerse themselves

© ICST Institute for Computer Sciences, Social Informatics and Telecommunications Engineering 2021
Published by Springer Nature Switzerland AG 2021. All Rights Reserved
E. I. Brooks et al. (Eds.): DLI 2020, LNICST 366, pp. 26–39, 2021.
https://doi.org/10.1007/978-3-030-78448-5_2

into learning and obtain better learning information. The field of DGBL has been widely researched and applied currently. Through this way of learning, it brings more pleasant and stimulating in the learning process which not only enhances motivation and interests, but also improves learning effectiveness.

DGBL is a valuable study for teachers and students in terms of teaching strategies and methods to enhance students' motivation, attitude and satisfaction, and thus enhance students' learning effectiveness. Therefore, this study will explore the relationship between the use of computer role-playing games in mythology literature, the usefulness of perception, the ease of perception, and the learning motivation, attitude, and satisfaction of students.

2 Reviewed Literature

2.1 Digital Game-Based Learning

Digital Game-Based Learning (DGBL) is a student-centered teaching method that can decrease the frustration of learners. It combines digital games with teaching content and considers learner's learning and cognitive style. Learning activities by simulating states can make the learning process more stimulating and effective [1]. DGBL can also be applied on Inquiry-based learning. The exploratory way of learning is an effective teaching strategy which provides students with the opportunity to explore in the game environment with tasks and challenges. It also allows learners to continually retry tasks and solve problems based on feedbacks from the game system. Inquiry-based learning integrates teaching strategies into the game environment and provides challenges in the game tasks. It can also arrange to elevate significant inquiry activities which can effectively improve learners' learning effectiveness, motivation, flow experience, learning satisfaction [2].

2.2 Learning Motivation

Motivation is the requirement to reach reserved goals and the initial point for a series of physiological processes that drive individuals to influence behavior [3]. How to stimulate students' learning motivation is one of the important topics for teachers in the classrooms. Motivation can be divided into internal and external forms according to the different causes or goals that are caused by the action. The intrinsic motivation refers to the way of internal reward rather than any external reward. The motivation for learning in this study is from the perspective of internal motivation. The origin of internal motivation is the personal desire, but not external reward or punishment [4]. It is the desire of students to continuously pursue knowledge and skills. Intrinsic motivation is motivated by interest or entertainment, even because of the internal desire to challenge and solve the target tasks [5]. As the learners acquire abilities and knowledge, intrinsic motivation leads to satisfaction and pleasure in the learning process [6]. To compared with extrinsically motivated students, intrinsically motivated students are more likely to persevere and maintain long-term learning motivation when facing the learning challenges [7].

2.3 Learning Attitude

Attitude can be a relatively stable cognitive and emotional psychological tendency expressed in a situation or concept. Attitude is different from motivation. Attitude is a group of beliefs, and motivation is the reason for doing something [8]. Attitudes are also usually different from beliefs, because attitudes have a moderate duration, strength, and stability. And they have emotional content as well, while beliefs are stable and difficult to change ideas [9]. Emotional factors such as attitude, motivation, and anxiety, play an important role in stimulating and supporting learning effectiveness [10]. In the study of higher education, regardless of the teaching method adopted by teachers in the college, it points out that the student participation may be the most important factor in determining whether college students can successfully learn or not [11]. Research on attitudes regarding the use of Information and Communication Technology has shown that there is greater potential for the use of digital learning technology in order to enhance learning activities. It can not only promote the participation in learning, but also enhance the positive learning attitudes [12].

2.4 Learning Satisfaction

Satisfaction is a kind of psychological sense and an abstract term as well. Martin [13] believes that satisfaction refers to the consistency between an individual's expectation of gaining experience and the actual results one feels when experiencing. If it equals or exceeds to expectations, one will feel satisfied. Otherwise, one will not be satisfied. In this way, learning satisfaction can be the degree of psychological expectation that the learner feels whether the learning activities are encountered. When the learning topics help to fulfill student expectations and needs, they can increase students' satisfaction. Kuo et al. [14] studied some predictive indicators of student satisfaction and pointed out that when learners and instructors or learners and learning materials interacted well, they were good predictors of learning satisfaction. Learning through a series of interactions allows learners and instructors or learners and textbooks to have a large number of good interactions, which is also a way to improve learning satisfaction. The higher learning satisfaction will also affect learning motivation and effectiveness as well [15].

3 Methodology

3.1 Teaching Material Content

This study uses the Input-Process-Outcome Game Model (IPOGM) [16]. It produces a computer role-playing game with the learning content in the game which is set to the Shan Hai Jing mythology literature course. The learner can control the roles in the adventure and puzzle solving of the game, under the rich and reasonable storyline. The learners can also enjoy in the game loop, and solve problems by exploring the environment, collecting information, and thinking strategies in order to achieve high engagement of participation in learning activities. The dialogue mode in the game is shown in Fig. 1, and the battle mode in the game is shown in Fig. 2.

Fig. 1. Game dialogue mode.

Fig. 2. Game battle mode.

3.2 Research Hypothesis

This study focuses on the relationship between the use of computer role-playing games in the teaching of mythology and literature courses. And it explores the relationships among perception of students' usefulness, ease of use, learning motivation, learning attitude, and learning satisfaction.

H1: Usefulness affects students' learning motivation.
H2: Ease of use affects students' learning motivation.
H3: Usefulness affects students' learning attitude.
H4: Ease of use affects students' learning attitude.

H5: Usefulness affects students' learning satisfaction.
H6: Ease of use affects students' learning satisfaction.

3.3   The Definition and Measurement of the Research Dimensions

All Dimensions of this study refer to relevant literature for the definition and operation of variables. As for the question of all the Dimensions, the "usefulness", "ease of use", and "learning motivation" scales were modified from Hwang et al. [17]. And the questions of the "learning attitude" scale were modified from Pierce et al. [18]. The questions on the "learning satisfaction" scale were modified from Sun et al. [19]. The questionnaire items and scale reference sources are shown in Table 1, Table 2, Table 3, Table 4 and Table 5.

Table 1. Questionnaire items and source.

Dimensions	Questions	Sources
Usefulness	I think that the use of computer role-playing games enriches learning activities in mythology literature courses	Hwang et al. [17]
	I think that the mythology literature courses using computer role-playing games is very helpful for me to gain new knowledge	
	I think that in mythology literature courses, the learning mechanism provided by computer role-playing games makes the learning process smoother	
	I think, in mythology literature courses, using computer role-playing games helps me getting valuable information when I need it	
	I think that in mythology literature courses, using computer role-playing games helps me learning better	
	I think it is more useful to apply computer role-playing games in mythology literature courses	

Table 2. Questionnaire items and source.

Dimensions	Questions	Sources
Ease of use	I think it is not difficult for me to use computer role-playing games in mythology courses	Hwang et al. [17]
	It only took me a short time to fully understand how to use computer role-playing games in mythology literature courses	
	I think that in mythology literature courses, learning activities using computer role-playing games are easy to understand and follow	

(*continued*)

Table 2. (*continued*)

Dimensions	Questions	Sources
	I quickly learned how to use computer role-playing games in mythology and literature courses	
	I think it is not difficult for me to use computer role-playing game learning methods in mythology literature learning activities	
	I think it is easy to use computer role-playing games in mythology courses	

Table 3. Questionnaire items and source.

Dimensions	Questions	Sources
Learning motivation	I think it is interesting to use computer role-playing games to learn mythology and literature courses	Hwang et al. [17]
	I think it is valuable to use computer role-playing games to learn mythology and literature courses	
	I want to use computer role-playing games to learn more in mythology and literature courses	
	I think it is worthy to use computer role-playing games to learn the content of mythology and literature courses	
	To me, learning mythology and literature courses is very important	
	I know that learning mythology and literature courses is very important for future applications	
	I will actively look for more information to learn mythology and literature courses	
	I think it is very important to learn mythology and literature courses for every student	

Table 4. Questionnaire items and source.

Dimensions	Questions	Sources
Learning attitude	I like to use computer role-playing games for mythology and literature courses	Pierce et al. [18]
	I think I will learn more when using computer role-playing games in mythology courses	

(*continued*)

Table 4. (*continued*)

Dimensions	Questions	Sources
	I think the use of computer role-playing games in mythology literature courses is worth to pay extra effort	
	I think mythology and literature courses are more interesting when using computer role-playing games	
	I think computer role-playing games can help me learning mythology and literature courses better	

Table 5. Questionnaire items and source.

Dimensions	Questions	Sources
Learning satisfaction	I am satisfied with the decision of using computer role-playing games to learn mythology and literature courses	Sun et al. [19]
	If have the opportunity to use computer role-playing games to learn mythology and literature courses, I would be happy to do so	
	I selected to use computer role-playing games to learn mythology and literature courses	
	I think to use the computer role-playing games to assist the mythology course is very satisfied	
	I think that mythology literature courses assisted with computer role-playing games satisfies my learning needs well	
	I will do my best to use computer role-playing games to learn mythology and literature courses	

3.4 Data Collection

The Likert's 5-point scale is used in this research. After taking an 80-min computer role-playing game in mythology and literature courses for the freshmen students of the Department of Digital Media Design of Ling Tung University, the questionnaire is distributed and filled out. A total of 108 copies of effective questionnaires were collected 75 females and 33 males were included in the questionnaire. The SPSS 21.0 is used as the computer statistical analysis software. The statistical analysis includes reliability analysis and the validity is tested through KMO. Also the research hypothesis has been analyzed. Through correlation analysis, we can understand the relationships between Dimensions. Through regression analysis, it provides further understanding the degree of mutual influence of Dimensions.

4 Research Results

4.1 Reliability Analysis

The interviewing questionnaire of this research includes five dimensions: "usefulness", "ease of use", "learning motivation", "learning attitude" and "learning satisfaction". According to Guielford [20], a Cronbach's alpha value greater than 0.7 indicates a high degree of confidence. In this study, 108 questionnaires were distributed, and 108 valid questionnaires were recovered, with an effective recovery rate of 100%. Cronbach's alpha values of the five dimensions, such as usefulness, ease of use, learning motivation, learning attitude, learning satisfaction, are higher than 0.8 (between 0.862–0.958). It possesses relatively high reliability. The five dimensions of Cronbach's alpha are listed in Table 6:

Table 6. Reliability analysis table.

Dimensions	Number of questions	Cronbach's alpha value
Usefulness	6	0.958
Ease of use	6	0.862
Learning motivation	8	0.936
Learning attitude	5	0.914
Learning satisfaction	6	0.946

4.2 Validity Analysis

In the validity part, the issuance questionnaires are supervised by three relevant experts with experience in designing questionnaires in order to complete the questionnaire design, so that the questionnaires have certain content validity.

Through the KMO Spherical Test. The KMO value is between 0 and 1. The closer the value is to 1, the higher the correlation of the variable is, which is more suitable for factor analysis. The closer the value is to 0, the lower the correlation of the variable is which is less suitable for factor analysis. The KMO measure of sampling adequacy is commonly used metrics above 0.9 indicate that it is very suitable; 0.8 indicates that it is suitable; 0.7 indicates that it is generally suitable; 0.6 indicates that it is not suitable; 0.5 or less indicates that it is extremely unsuitable. The overall test results of this study are summarized in the following Table 7:

Table 7. The KMO spherical test.

Dimensions	KMO Value	Results
Usefulness	0.908**	Very suitable
Ease of use	0.759**	Generally suitable
Learning motivation	0.851**	Suitable
Learning attitude	0.881**	Suitable
Learning satisfaction	0.894**	Suitable

** significant at $p < 0.01$

Among the dimensions, the KMO value of learning motivation is 0.851, which passes the verification standard. The interpretable variation is 69.46% which can be extracted of a component. The KMO value of learning attitude is 0.881, which passes the verification standard. The interpretable variation is 74.43% which can be extracted of a component. The KMO value of learning satisfaction is 0.894, which passes the verification standard. The interpretable variability is 79.28% which can be extracted of a component. The KMO value of ease of use is 0.759, which passes the verification standard and can explain 59.30% of the variation. It can be extracted a component. The KMO value of usefulness is 0.908, which passes the verification standard. The explainable variation is 86.60% which can be extracted of a component. The overall test results of this study are summarized in the following Table 8. Therefore, all the dimensions of this study have passed the standard with certain validity.

Table 8. The KMO spherical test.

Dimensions	KMO Value	Explanation Total Variance
Usefulness	0.908**	86.60%
Ease of use	0.759**	59.30%
Learning motivation	0.851**	69.46%
Learning attitude	0.881**	74.43%
Learning satisfaction	0.894**	79.28%

** significant at $p < 0.01$

4.3   Correlation Analysis

In order to understand the influence variables between the facets and whether there is a relationship between the facets, the results are shown in Table 9 through Pearson correlation analysis:

Table 9. Correlation analysis table.

Dimensions	Usefulness	Ease of use	Learning motivation	Learning attitude	Learning satisfaction
Usefulness	1				
Ease of use	0.71**	1			
Learning motivation	0.81**	0.61**	1		
Learning attitude	0.87**	0.69**	0.86**	1	
Learning satisfaction	0.91**	0.67**	0.84**	0.91**	1

** significant at p < 0.01

As shown in Table 9, there are correlations between the facets. This study will further carry out regression analysis.

4.4   Regression Analysis

In order to understand the influence among the facets, this study will carry out regression analysis to clarify the relationship between each other. It used usefulness and ease of use as independent variables, and respectively analyzed learning motivation, learning attitude, and learning satisfaction as dependent variables. The result are summarized in Table 10:

Table 10. Table of three-mode regression analysis arrangement.

Independent variable/dependent variable	Model 1: learning motivation	Model 2: learning attitude	Model 3: learning satisfaction
Usefulness	0.759**(p = 0.000)	0.803**(p = 0.000)	0.603**(p = 0.000)
Ease of use	0.076 (p = 0.343)	0.096 (p = 0.065)	0.046 (p = 0.418)
Significance	0.000**	0.000**	0.000**
Degree of freedom	2	2	2
F value	103.605	199.100	262.630
R^2	0.664	0.791	0.833

** significant at p < 0.01

In Model 1, we use learning motivation as a dependent variable, through usefulness and ease of use as independent variables. We found that the model is established and the regression analysis was found that usefulness affects the learner's learning motivation (0.759**). But the perception ease of use has not affected.

In Model 2, we use learning attitude as a dependent variable, through usefulness and ease of use as independent variables. We found that the model is established and the regression analysis found that usefulness affects the learner's learning attitude (0.803**). But the perception ease of use has not affected.

In Model 3, we use learning satisfaction as a dependent variable, through usefulness and ease of use as independent variables. We found that the model is established and the regression analysis found that usefulness affects learner's learning satisfaction (0.603**). But the perception ease of use has not affected.

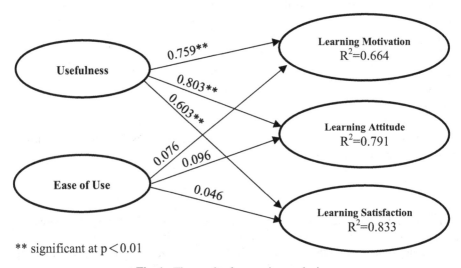

** significant at $p < 0.01$

Fig. 1. The result of regression analysis.

4.5 Summary

Through the statistical analysis of the data from above, in addition to having good relia-bility and validity, this research has an impact on the relationship between "usefulness", "learning motivation", "learning attitude", and "learning satisfaction". However, "ease of use" is invalid for "learning motivation", "learning attitude" and "learning satisfac-tion". The inference may be because students are now fairly familiar with the operation of computer role-playing games. Therefore, the ease of use is not an important con-sideration. However, in terms of teaching strategies, teachers must pay attention to the usefulness of learners' perception. If the learners' perception is useful, it can greatly increase the chances of achieving success in teaching. And it can make students have more comparable learning motivation, attitude, and satisfaction. Based on the above, the results of the hypothesis in this study are summarized in Table 11:

Table 11. The hypothesis of the establishment.

Hypothesis	Status
H1: Usefulness influences the learning motivation of computer role-playing games in teaching mythology and literature	True
H2: Ease of use influences the learning motivation of computer role-playing games in teaching mythology and literature	False
H3: Usefulness influences the learning attitude of computer role-playing games in teaching mythology and literature	True
H4: Ease of use influences the learning attitude of computer role-playing games in teaching mythology and literature	False
H5: Usefulness influences the learning satisfaction of computer role-playing games in teaching mythology and literature	True
H6: Ease of use influences the learning satisfaction of computer role-playing games in teaching mythology and literature	False

5   Conclusion

5.1   Research Finding

(1) Usefulness will positively affect students' learning motivation (H1)
The research confirms that computer role-playing games are used to teach mythology and literature courses. Usefulness will positively affect students' learning motivation. Therefore, the hypothesis H1 of this study is confirmed.
(2) Ease of use cannot positively affect students' learning motivation (H2)
The research confirms that computer role-playing games are used to teach mythology and literature courses. Ease of use does not positively affect students' learning motivation. Therefore, the hypothesis H2 of this study is not confirmed.
(3) Usefulness will positively affect students' learning attitude (H3)
The research confirms that computer role-playing games are used to teach mythology and literature courses. Usefulness will positively affect students' learning attitudes. Therefore, the hypothesis H3 of this study is confirmed.
(4) Ease of use cannot positively affect students' learning attitude (H4)
The research confirms that computer role-playing games are used to teach mythology and literature courses. Ease of use cannot positively affect students' learning attitudes. Therefore, the hypothesis H4 in this study is not confirmed.
(5) Usefulness will positively affect students' learning satisfaction (H5)
The research confirms that computer role-playing games are used to teach mythology and literature courses. Usefulness will positively affect students' learning satisfaction. Therefore, the hypothesis H5 of this study is confirmed.
(6) Ease of use cannot positively affect students' learning satisfaction (H6)
The research confirms that computer role-playing games are used to teach mythology and literature courses. Ease of use cannot positively affect students' learning satisfaction. Therefore, the hypothesis H6 of this study is not confirmed.

5.2 Research Conclusion

In addition to the nice reliability and validity of this research, when discussing the impact relationship, "Usefulness" has a respective effect on "learning motivation", "learning attitude", and "learning satisfaction". However, "Ease of use" does not hold for "learning motivation", "learning attitude", and "learning satisfaction". This result is somewhat different from the research of some scholars [21]. The inference of this research is that students in the Department of Digital Content Design have a high chance of using computers. In addition, computer role-playing games are so popular in Taiwan that students are very familiar with the operation of such computer role-playing games. Therefore, ease of use is no longer an important consideration for students. However, in terms of teaching strategies, teachers should pay attention to the usefulness of learners' perception. If the learners' perception is useful, it can greatly increase the chances of success in teaching. It can also make students have certain promoted degree of comparable learning motivation, attitude, and satisfaction.

References

1. Hung, P.H., Hwang, G.J., Lee, Y.H., Su, I.H.: A cognitive component analysis approach for developing game-based spatial learning tools. Comput. Educ. **59**(2), 762–773 (2012)
2. Hwang, G.J., Chiu, L.Y., Chen, C.H.: A contextual game-based learning approach to improving students' inquiry-based learning performance in social studies courses. Comput. Educ. **81**, 13–25 (2015)
3. Moos, D.C., Marroquin, E.: Multimedia, hypermedia, and hypertext: motivation considered and reconsidered. Comput. Hum. Behav. **26**(3), 265–276 (2010)
4. Deci, E.L., Ryan, R.M.: The "what" and "why" of goal pursuits: human needs and the self-determination of behavior. Psychol. Inq. **11**(4), 227–268 (2000)
5. Martens, R.L., Gulikers, J., Bastiaens, T.: The impact of intrinsic motivation on e-learning in authentic computer tasks. J. Comput. Assist. Learn. **20**, 368–376 (2004)
6. Elliot, A.J., Harackiewicz, J.M.: Approach and avoidance achievement goals and intrinsic motivation: a mediational analysis. J. Pers. Soc. Psychol. **70**, 461–475 (1996)
7. Huang, W.D.: Evaluating learners' motivational and cognitive processing in an online game-based learning environment. Comput. Hum. Behav. **27**(2), 694–704 (2011)
8. Oroujlou, N., Vahedi, M.: Motivation, attitude, and language learning. Procedia. Soc. Behav. Sci. **29**, 994–1000 (2011)
9. Mayes, R.: ACT in algebra: student attitude and belief. Int. J. Comput. Algebra Math. Educ. **5**(1), 3–13 (1998)
10. Lin, H., Hong, Z.R., Chen, Y.C.: Exploring the development of college students' situational interest in learning science. Int. J. Sci. Educ. **35**(13), 2152–2173 (2013)
11. Trowler, V., Trowler, P.: Student Engagement Evidence Summary. Higher Education Academy (2010)
12. Ciampa, K.: Learning in a mobile age: an investigation of student motivation. J. Comput. Assist. Learn. **30**(1), 82–96 (2014)
13. Martin, C.L.: Enhancing children's satisfaction and participation: using a predictive regression model of bowling performance norms. Phys. Educ. **45**(4), 196–209 (1988)
14. Kuo, Y.C., Walker, A.E., Belland, B.R., Schroder, K.E.: A predictive study of student satisfaction in online education programs. Int. Rev. Res. Open Dist. Learn. **14**(1), 16–39 (2013)

15. Lovecchio, C.P., DiMattio, M.J.K., Hudacek, S.: Predictors of undergraduate nursing student satisfaction with clinical learning environment: a secondary analysis. Nurs. Educ. Perspect. **36**(4), 252–254 (2015)
16. Garris, R., Ahlers, R., Driskell, J.E.: Games, motivation, and learning: a research and prac-tice model. Simul. Gaming **33**(4), 441–467 (2002)
17. Hwang, G.J., Yang, L.H., Wang, S.Y.: A concept map-embedded educational computer game for improving students' learning performance in natural science courses. Comput. Educ. **69**, 121–130 (2013)
18. Pierce, R., Stacey, K., Barkatsas, A.: A scale for monitoring students' attitudes to learning mathematics with technology. Comput. Educ. **48**, 285–300 (2007)
19. Sun, P.C., Tsai, R.J., Finger, G., Chen, Y.Y., Yeh, D.: What drives a successful e-Learning? An empirical investigation of the critical factors influencing learner satisfaction. Comput. Educ. **50**(4), 1183–1202 (2008)
20. Guielford, J.P.: Fundamental Statistics in Psychologyand Education, 5th edn. McGraw-Hill, New York (1973)
21. Bourgonjon, J., Valcke, M., Soetaert, R., Schellens, T.: Students' perceptions about the use of video games in the classroom. Comput. Educ. **54**, 1145–1156 (2010)

Potentiating Digital Educational Environments Through Data Analytics

Flávio Lima Faria[1]([✉]) [iD], Maitê Gil[2] [iD], Eva Oliveira[1] [iD], and Cristina Sylla[3] [iD]

[1] 2Ai, Polytechnic Institute of Cávado and Avenue, Barcelos, Portugal
[2] Research Centre on Child Studies, Universidade do Minho, Braga, Portugal
[3] Research Centre on Child Studies – ITI/LARSyS, Guimarães, Portugal

Abstract. When we think of educational games or apps, data analysis is not what comes first to our mind. However, a fundamental feature of all types of games/apps is the data. When playing, the system is constantly collecting data from the users, about the current state of the game and making predictions and decisions based on that data. This is what data analytics is all about. Games are an amazing way for scientists and educators to communicate with multiple data sets, as well as a great way for developers to get a wealth of relevant information to inspire the development of new algorithms and theories. The main goal of this investigation is to propose and validate techniques that allow the application of data analysis concepts to improve digital educational environments, especially the ones related to storytelling. In this paper, we discuss the development and implementation process of specific analytical techniques applied to Mobeybou, a set of story apps for children, as a first step to develop a set of guidelines to support, inform and optimize the evaluation of the efficiency of educational games/apps using data analysis.

Keywords: Data analytics · Educational games · Interactive story apps · Children

1 Introduction

This work aims to develop a set of guidelines and techniques to optimize the evaluation of the efficiency of educational tools, such as games and story apps directed to children using data analysis. The goal is not to develop a new data mining or analytics tool, but rather to propose and validate techniques that allow the application of data analysis concepts to improve digital educational environments, especially the ones related to storytelling. With these techniques, we intend, for example, (i) to collect and analyse relevant data of the users and their contexts, in order to understand and optimize the learning process in environments that use educational games, and (ii) to better understand students' learning process. In order to develop this general framework, we started by customizing existing data collection methods and algorithms that we have then implemented in the Mobeybou story apps [1]. In this paper, we present and discuss the development and implementation process of these specific analytical techniques.

E. I. Brooks et al. (Eds.): DLI 2020, LNICST 366, pp. 40–47, 2021.
https://doi.org/10.1007/978-3-030-78448-5_3

2 Data Analytics and Games

Any team responsible for designing a game needs to understand its users, e.g., why do they play, why do they stop playing, how long do they play? These questions can be answered by analyzing data collected during the game play and thus informed design decisions can be made instead of relying on mere assumptions. By integrating data collection into a game, it is possible to collect different types of information about the users: How many people play the game every day, how many times do these people play during a certain period of time, what stages of the game do they play the most or how long do they spend playing each session. In short, data analysis allows designers to focus on the most relevant tasks to improve their games and to make them more attractive to their users [2].

2.1 Data Analytics and Educational Games/Apps

Regarding educational contexts, Agasisti and Bowers (2017) propose that the instruments to be used for data analysis in specific research and to answer the desired questions can be ordered in three different approaches considering its specific purpose: Educational Data Mining (EDM), Learning Analytics (LA) and Academic Analytics (AcAn).

Educational Data Mining uses computational methods to detect patterns and recurrences in large amounts of collected educational data. The main beneficiaries of this approach are researchers, analysts, faculty and tutors. EDM represents a host of education-specific machine learning tools, providing an excellent foundation for our discussion of learning design analytics. EDM is an emerging discipline, concerned with developing methods for exploring large educational data streams [3].

Learning Analytics uses many of the Educational Data Mining analysis models but focuses on teaching and learning activities with great attention to the achievement goals, that is, the results. LA uses data analysis to help researchers understand the factors that are critical to learning. Learners, faculty and tutors can benefit from this data analysis method.

Academic Analytics, on the other hand, focuses on the systematic use of data to generate information that can improve the internal efficiency of operations and management processes in educational institutions. Thus, its use is more focused on the organizational level of education, such as personnel, resources and material management. In our work, we intend to use tools from the first two approaches (EDM and LA) to assist our research, as we believe they are more relevant to analyze the data collected from the Mobeybou users in order to improve the quality of storytelling and the educational aspects of the app.

When applied during the various stages of game development, data analysis methods can provide in-depth insights of learning patterns in digital game environments, as well as inform and facilitate design iterations for optimizing learning and engagement [3]. These insights can inform each stage of the game design and production and be carried out in synchrony with game refinement cycles to fuel iterative, data-driven design for better engagement in learning. In the early stages of development, the definition of the data structure and visualization analysis can be valuable in supporting design, since the discovery of data structures can reveal deeper interaction patterns performed by the user as the mechanics are solidified in the following stages of development. This predictive approach can support game production by providing learning prediction and behavior detection for adaptability in the game supporting the learning objectives [4].

There are several methods for evaluating in-game learning: "Broad analysis goals (or 'metagoals') common to the expert EDM synopses are visualization, relationship mining, and prediction [...]. Visualization involves graphic representations of data; relationship mining looks at associative data patterns; and prediction can project outcomes via algorithms of sequence, probability, and regression. Specific method types include: descriptive visualization, social networks, clustering, association, classification/regression, and pattern mining"[5:2].

A simple example is to measure how many levels a student has completed or to record the last completed level. The benefit of this approach is that it is very straightforward and relatively easy to implement. The analysis of the last challenge a student successfully overcame can be used as an independent assessment variable of the student's performance [6]. Andersen et al. [7] applied a different method for evaluating success in a study of the effects of different tutorial styles on learning gameplay, looking at how well a player was learning while playing a game. This approach works well when the gameplay mechanics are properly aligned with the learning content.

Another method is Evidence Centered Design (ECD) [8], which depends on three main components: a competency model, an evidence, and a task model. The competency model describes the knowledge level intended to be achieved with the system. The evidence model analyzes student's interactions with a system through predefined rules. The evidence model also shows a relationship between the competence model and the observed scores. Finally, the task model provides a framework for describing the tasks a student performs within the system [9]. In this investigation, we focus only on the post-game data collection method, that is, the data is analyzed after the sessions and not in real time. The data collected in the app used as a model is grouped and made available for analyses at 12-h intervals.

This brief panorama of evaluation methods for educational games indicates a diversity of practices. The fact that educational games have been used as an innovative instructional strategy to make learning more effective justifies such diversity, since it is important to systematically evaluate these games and check for evidence of its impact on learning. However, a systematic literature review carried out by Petri and von Wangenheim [10], in which the authors identify and describe seven different approaches to systematically evaluate educational games, has shown that most of the approaches are developed in an ad-hoc manner and do not provide an explicit definition of the study, its execution and data analysis. Therefore, the authors argue that there is a need for research on the definition and operationalization of educational game evaluations. According to Steiner et al.: "A crucial question is how to harness and make sense of game-based user data in an educationally relevant manner" [11:196]. In this sense, the data analytics techniques developed in this study are an effort to obtain more systematic data collection and, consequently, more valid and uniform results.

Although evidence of learning is the main objective when investigating, developing or designing educational games, it is not the only aspect that can be investigated on a game. Many other important questions may arise during the development of an educational game. Among them, it is important to highlight the following: Are the goals of the game really rewarding the kinds of behaviors that the game wants to encourage? Does the difficulty of the levels progress adequately to the proposed objectives? Answering these questions may affect the interpretation of learning measures and may have implications for the redesign of game mechanics. Gameplay data analysis can be a valuable ally in determining what types of mechanics work, or not, to meet the learning goals. However, the obtained data can only be consistently and assertively analyzed when these learning goals are well defined [12].

The use of data analytics in educational environments can inform both educators and designers. For educators, the data collection can be seen as an iterative cycle of hypothesis formation, testing and improvement. E.g., the educator can understand which learning topics need to be reinforced. In this process, the objective is not only to transform data into knowledge, but also to apply the extracted knowledge to make decisions about how to modify the digital educational environment to improve student's learning. For game designers, data analytics is a way to evaluate the design of applications or games and can help to improve the game design. Therefore, data collection in educational environments can be a powerful ally for assessing the effectiveness of interactive experiences [13].

Our literature review identified a lack of detailed work on data analysis in educational games. This review also indicates that there is a growing interest in evidence-based education, where games and educational applications that aim to improve teaching and learning can be validated with real data obtained from user interactions.

3 Developing Data Analytics Techniques for the Mobeybou Interactive Story Apps

Based on the assumptions presented above, we propose the development of a set of data analytics techniques to be integrated in the Mobeybou apps, a set of interactive story apps directed to pre and primary school children. The integration of these techniques will allow the collection of relevant data of the user's interactions with the story apps that will inform (i) future design decisions and (ii) the evaluation of the user's learning process. In this section, we describe the decisions regarding the development of such data analytics techniques.

The development of techniques for capturing the user's activity data has great relevance for validating the collected information. These techniques should be concise and built to collect and catalog the data in a reliable manner. Based on the above presented understandings, we developed an interaction logging system, a kind of activity map, which, by saving the user's activity log, can provide a database to inform teachers and researchers about which game mechanics and elements are more effective in promoting the user's achievement of the learning objectives.

Several methodological procedures are necessary to accomplish this goal, namely: (i) choosing the data platform in which the study is going to be developed; (ii) defining which data will be collected and recorded considering the learning goals; (iii) customizing the data analytics techniques; (iv) implementing the data analytics techniques; (v) establishing a time window for the collection of this data; (vi) analysing the collected data; and, finally, (vii) indicating, based on the analysis of the data, paths for both the improvement of the story apps and the development of a set of guidelines and techniques for using data analysis in educational applications or games with a storytelling component.

3.1 Using Unity Analytic as a Data Platform

We choose the Unity Analytics data platform to carry out this investigation since it is a simple but powerful data platform that also provides data recording for projects developed using Unity. Unity Analytics, is the only analysis solution integrated in the Unity engine that does not require the installation of a third-party SDK (Software Development Kit). Unity Analytics allows collecting specific information that helps to adjust the gameplay and offer the best possible experience to the users. To obtain the behavioral data it is necessary to record the events or activities that the user performs within the game. Specific actions or choices carried out by the users are captured. To do this, a series of events are created that are triggered by the user's actions and indicate Unity Analytics to capture certain values. The collected data is generally divided into the following categories: geodemographic data; behavioral data; operational data; funnels; and qualitative data [14].

Along the entire data collection process it is mandatory to assure the anonymity of the users. Therefore, the data collection is merely used, to understand how the game applies to all users, and only relevant data is stored. Some types of data are irrelevant and some data can also be inferred by evaluating interconnected events. Understanding the kinds of data we can access and its meaning is fundamental to convert the collected data into improvements or new features for a project. In the next section, we present the decisions we have made about which data needs to be collected and stored by the data analytics techniques under development.

3.2 Design of Adaptive Data Analysis Techniques

In order to collect relevant data using the Unity Analytics platform, we apply scripts at key points in the story app that serves as a test model. These scripts act as a trigger that records the user's actions within the interactive application.

The first design decision was to map every button of the app's opening interface. The recording of all possible interactions in the opening interface is necessary to get information about how the user navigates through the user interface. Important users' choices within the opening interface that have pedagogical implications are, for instance, related to the following questions: What does the player do first: Read the story, read the glossary or play the integrated game? Does s/he read the story from the beginning or selects a different page to begin the reading? Which page do the users read most frequently? Which page do the users read less frequently? Which sequence of pages do they prefer to read? How many users visit the glossary page?

This kind of data can help researchers and educators to get an overview of the users' reading behaviour, which is relevant both to improve the interactive story application - for instance by redesigning the less visited pages, - and to develop pedagogical approaches

- for example, by focusing on the preferred sequence of pages or on the integration of the glossary into the story.

The second design decision was to map every button of the app's menu interface. This is relevant to record the most selected language or the most chosen character by the users, as well as to access if the users turn off the music or the integrated narration of the story. In the story pages, the data is collected each time the users touch an interaction area and the navigation buttons. The analysis of such interconnected events allows us to infer valuable information such as: How long does it take users to read each page? How many users give up on completing the reading? Do the users do all possible interactions before moving to the next page? What kind of interaction seems to involve users the most? Again, interpreting such data can lead to relevant understandings about the users' reading behaviour and inform design and pedagogical decisions.

We also map data such as the user's life cycle, how many users are online, how many use the app daily and how many times each user interacts with the app. The country in which the user is located is also registered.

The above mentioned aspects are examples of custom events that we use to capture data of the player's behavior, however, and depending on the defined goals, parameters are often changed over time, some other parameters can be added and others removed. This is why simplicity and ease of implementation and management for data analysis techniques for educational games or applications is so important. In our investigation, the developed data analytics techniques were implemented in May 2020, and will collect data during a time window of six months. After that, the collected data will be analysed to understand if the chosen parameters and custom events are effective for informing future design and pedagogical decisions, as well as good enough to support the development of a technique for using data analysis techniques for educational applications or games with a storytelling component.

Concerning the analysis of the collected data, it is important to highlight that it requires new techniques, since interactive learning platforms, such as digital games, have opened up new opportunities to collect and analyze student's data, to map patterns and trends in that data and to test hypotheses about how students learn.

As discussed in Sect. 2.1, there is a lack of well-established and well-founded techniques to analyze the efficiency of educational games and applications. Data collection, for example, is performed in various ways and not all are clearly justified. Although different data may be collected, there is still little guarantee about which data is most suitable for the evaluation of educational games. However, if properly designed and applied, the use of data analytics can be a valuable contribution to studies that aim at evaluating educational games by operationalizing an important phase for the evaluation, that is, the collection of data.

4   Conclusion and Future Work

Along this investigation, we have found evidence that data analysis can be a valuable tool for evaluating educational games. Following the customization of data analytics techniques to implement in story apps for children, a set of guidelines to optimize the evaluation of the efficiency of educational games using data analysis will be collected.

Future work includes: (i) analysis of the data recorded during this investigation in order to determine if the selected data was adequate to answering the questions related to the story apps and their learning goals; (ii) the design of general techniques to determine which data is important to collect in order to answer the desired research questions; (iii) the design of data management techniques that can be adaptable and easily implemented in educational game projects. As a long-term goal, we plan to apply data analysis in educational games in an agile and adaptable way with the aim of providing data that can help to improve teaching in educational games.

Acknowledgments. MoBeyBOU: Moving Beyond Boundaries - Designing Narrative Learning in the Digital Era, has been financed by national funds through the Portuguese Foundation for Science and Technology (FCT)- and by the European Regional Development Fund (ERDF) through the Competitiveness and Internationalisation Operational Program under the reference POCI/01/0145/FEDER/032580.

References

1. Sylla, C., et al.: Designing narrative learning in the digital era. In: Proceedings of the ACM International Conference on Human Factors in Computing Systems, CHI 2019 Extended Abstracts, Glasgow, Scotland, UK, 4–9 May. ACM Press, New York (2019)
2. Tlili, A., Chang, M.: Data Analytics Approaches in Educational Games and Gamification Systems. Springer, Singapore (2019). https://doi.org/10.1007/978-981-32-9335-9. ISBN: 9813293357, 9789813293359
3. Freire, M., Serrano-Laguna, Á., Manero, B., Martinez-Ortiz, I., Moreno Ger, P., Fernández-Manjón, B.: Game Learning Analytics: Learning Analytics for Serious Games (2016). https://doi.org/10.1007/978-3-319-17727-4_21-1
4. Agasisti, T., Bowers, A.: Data Analytics and Decision-Making in Education: Towards the Educational Data Scientist as a Key Actor in Schools and Higher Education Institutions (2018)
5. Owen, V.E.: Using Learning Analytics to Support Educational Game Development: A Data-Driven Design Approach (2015)
6. Delacruz, G.C., Chung, G.K.W.K., Baker, E.L.: Validity Evidence for Games as Assessment Environments (CRESST Report 773), Los Angeles, CA, USA (2010)
7. Andersen, E., Rourke, E.O., Liu, Y., et al.: The impact of tutorials on games of varying complexity. s.l. : Proceedings of the CHI 2012, pp. 59–68 (2012)
8. Mislevy, R.J., Steinberg, L.S., Almond, R.G.: On the structure of educational assessments. Meas. Interdisc. Res. Persp. **1**, 3–62 (2003)
9. Shute, V.J.: Stealth assessment in computer-based games to support learning. In: Tobias, S., Fletcher, J.D. (eds.) Computer Games and Instruction. s.l. : Information Age Publishers, pp. 503–524 (2011)
10. Wangenheim, G.P., Gresse von, C.: How to Evaluate Educational Games: A Systematic Literature Review (2016)
11. Steiner, C., Kickmeier-Rust, M., Albert, D.: Making Sense of Game-based User Data: Learning Analytics in Applied Games (2015)
12. Wang, H., Sun, C.-T.: Game Reward Systems: Gaming Experiences and Social Meanings (2012)

13. Alonso-Fernandez, C., Calvo-Morata, A., Freire, M., Martinez-Ortiz, I., Fernández-Manjón, B.: Applications of data science to game learning analytics data: a systematic literature review. Comput. Educ. 141 (2019) https://doi.org/10.1016/j.compedu.2019.103612
14. Documentation, Unity scripting languages manual – Unity Blog. Unity Technologies Blog. https://docs.unity3d.com/Manual/UnityAnalytics.html

Virtual- and Augmented Reality-Supported Teaching for Professional Caregivers

Anders Kalsgaard Møller[(✉)] and Eva Brooks

Department of Culture and Learning, Aalborg University, 9200 Aalborg, Denmark
ankm@hum.aau.dk

Abstract. Virtual reality is used to support teaching in the professional caregiver educations. Virtual reality can visualize the anatomical part of the body and provide interactive content that support and motivate students. This paper presents a preliminary study of an ongoing development of two new virtual and augmented reality applications with teaching material about leg ulcer and Chronic Obstructive Pulmonary Disease. We have conducted workshops and interviews with a total of seven teachers at a caregiver program who have provided information about the potential uses of virtual and augmented reality including their design ideas and feedback on a prototype of the applications. This have contributed to insights into how future solutions should be designed and how these may support teaching. The results indicate that virtual and augmented reality can be useful for achieving many different learning goals. It also indicates that it is not always an advantage to use immersive virtual reality or augmented reality, but it is possible to achieve at least similar results using a non-immersive version through mobile phones or tablets.

Keywords: Virtual reality · Augmented reality · Learning · Caregivers · Health · Vocational training · Teaching

1 Introduction

Virtual reality (VR) is a technology that is being used more frequently lately in connection with education and training as it offers options for creating 3D spatial representations, multisensory channels for user interaction, immersion of the user and intuitive interaction through natural manipulations in real time [11]. One of the areas where VR has been used is in vocational training and often related to medicine and healthcare applications [8]. For example, to train surgeons [3], in nurse education [5], medical professionals [6] or for health care training [7]. It has also been used in rehabilitation therapy and training [4] and has been deemed effective for both mental- and physical therapy [10].

At the Danish Vocational Education and Training Colleges (VET), VR has been introduced to support teaching at the professional caregiver education. VR applications such as Sharecare VR [2] is used in anatomy courses to visualize the different parts of the body and other VR applications are used to introduce students to assistive technologies or simulate symptoms of dementia [9].

© ICST Institute for Computer Sciences, Social Informatics and Telecommunications Engineering 2021
Published by Springer Nature Switzerland AG 2021. All Rights Reserved
E. I. Brooks et al. (Eds.): DLI 2020, LNICST 366, pp. 48–60, 2021.
https://doi.org/10.1007/978-3-030-78448-5_4

Many of the students at the VET colleges have reading difficulties, dyslexia, attention-deficit/hyperactivity disorder (ADHD) and similar conditions. They have trouble with learning and keeping focus from traditional teaching methods such as lectures and textbooks. However, the introduction of VR has had a positive effect on the students learning outcome, which in turn have led to a desire to develop additional solutions to support teaching. Thus, plans have been made to develop two new applications targeting leg ulcers and Chronic Obstructive Pulmonary Disease (COPD). It is planned to develop both a VR version and an Augmented Reality (AR) version of the applications.

The difference between VR and AR is defined by Azuma (1997) who describe AR as a variation of VR where the difference is that "Virtual environment technologies completely immerse a user inside a synthetic environment.

While immersed, the user cannot see the real world around him. In contrast, AR allows the user to see the real world, with virtual objects superimposed upon or composited with the real world. Therefore, AR supplements reality, rather than completely replacing it [1].

One of the reasons for also designing an AR option is the option of using the application while still being able to participate in group work or other class-related activities. If a student is immersed in VR the student will not be able to see and interact with other group members or follow class-related activities.

In this study we hosted workshops and interviews with teachers at VET colleges around Denmark. Based on the input from the teachers we look at the opportunities and perspectives for VR-supported teaching at the caregiver education. More specifically, the study focus on what kind of learning goals VR can support, how these are supported and how future solutions should be designed.

2 VR and Learning

First-order experiences, with direct non-symbolic interaction, size, transduction, reification, autonomy, and presence are according to [1] what contributes to positive learning outcomes in VR. The sense of presence is defined by [13] as *"The sense of being there"* which enhances a user' first-hand experience where the user interacts directly with the world [11]. Transduction, size and reification are terms used to describe how raw data from the world is transformed into information we can interact with or experience through our senses e.g. enabling a user to interact with tissue or moving around inside a human body [12]. VR can support training in safe environments and increase learner's motivation [8]. It can promote active participation, high interactivity and individualization [14].

Some users do, however, have greater benefit from VR than others because the individual traits and characteristics influence the experience of the users in VR [20]. Not all users enjoy VR [15]. Some feels unsafe entering the virtual world and others suffer from cyber sickness.

A distinction can be made between VR/AR presenting a 3D world through a screen and a fully immersive experience presented through either a head mounted display (HMD) or a CAVE environment [16] where the virtual environment surrounds a person providing a fully immersive experience. The first is referred to as non-immersive VR/AR or desktop VR/AR and the latter as immersive VR/AR [8]. Results from [17] show that users are more engaged when using HMD compared to non-immersive alternatives and that they take it more seriously, e.g. approaching dangerous situations with more care [18]. Furthermore, users report being more present in immersive VR which is important for the learning process and make them spend more time in the training environment [12].

VR can be used for different types of skill acquisition both cognitive, psychomotor and affective skills [15]. In a literature review, [15] it was found that VR using HMD was better for learning visual and spatial information than alternative options. This could for instance be spatial information and visual information about the organs. However, it was found that non-HMD options were better for remembering facts such as names.

While not much evidence exist, studies indicated that HMD-based VR is superior when it comes to learning affective skills - as they: *"are related to interpersonal skills, and here the ability of the technology to create a believable simulation of a virtual human or social situation is crucial"* [15].

For psychomotor skills it was not important how the VR was presented. The learning outcome where closer related to the perceived realism and the quality of the kinesthetic input and haptic output provided through e.g. a joypad and how it fits with bodily movements [15].

3   The VR/AR Prototypes

A prototype has been created of the solution based on initial feedback from teachers and coordinators at the VET colleges. The prototype has been built as a VR solution for different versions of Oculus glasses and an AR version for Microsoft HoloLens. For the HoloLens version it is possible to interact using finger gestures, e.g. press a virtual button with a finger. For the VR version a joypad is used.

The prototype supports interaction and visualization of the different parts of the body with a special focus on the organs. The prototype provides basic information about the different parts of the body such as, names and information about how they function. Different options exist for interacting with different body parts. For instance, simulate different types of illness, turn the 3D model around or divide the body parts into different layers. In Fig. 1, the main menu is presented where one can select different parts of the body for further interactions and visualizations. Once an item is selected in the menu a 3D model of the item is presented. In Fig. 2, the lungs have been selected and a screen with a 3D model of the lungs is presented along with an information box. The user can turn the 3D model and select different stages of COPD. The user can also select different layers of the lungs (Fig. 3). Figure 4 shows a leg ulcer with an option to simulate different stages of the illness.

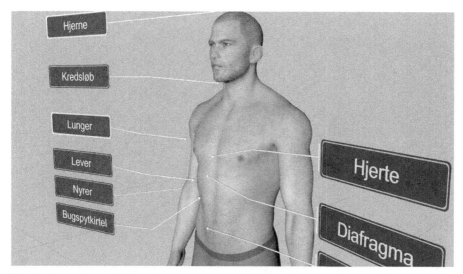

Fig. 1. The main menu of the AR/VR application where you can select different bodyparts/organs for further interactions.

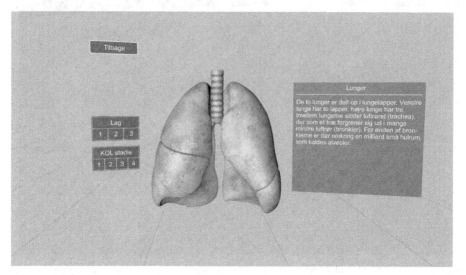

Fig. 2. 3D model of the lungs, a box with information about the lungs and buttons to select different stages of COPD and options for changing the layers this view is layer 1.

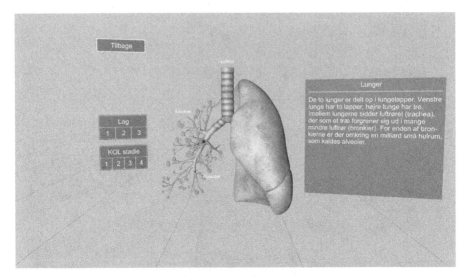

Fig. 3. 3D model of the lung with layer 3 selected where the bronchial and alveoli are visual.

Fig. 4. A model of a leg ulcer where one can change the severity of the leg ulcer by pulling the lever.

4 Method

It was originally planned to carry out a workshop with two groups of four teachers from the caregiver education. However, due to changed restrictions with Covid19 which occurred in the middle of the course, it instead became a workshop with three teachers and four single interviews with teachers. In total seven teachers participated in the study. The interviews were held online using Microsoft Teams. The teachers were all females and educated nurses. They all taught in somatic illness, anatomy and pharmacology and had between 1–12 years teaching experience.

The workshop consisted of a group interview with questions regarding their teaching experiences, knowledge of VR/AR, how they currently taught classes about COPD and leg ulcers, and how they considered VR/AR to be used to support their teaching activities. They were also asked to draw and explain their best idea for a VR/AR solution that could be used to support their teaching. Finally, they were introduced to the prototype of the new applications and given a chance to comment on the content, design and options for interaction. The prototypes were introduced in the end to ensure an open-minded approach where they could come up with new ideas that were not affected by the knowledge of the prototypes.

The single interviews followed the same procedure as the workshop, except that they did not get the option of drawing the solution. Instead of experiencing the prototype first-hand they were instead presented with a video of the prototype.

The workshop was recorded with a camera and the online interviews recorded using the video-recording option in Teams. The workshop and the interviews were transcribed verbatim and analyzed with a thematized approach [19] resulting in three themes: 1) Students and motivation, 2) Learning activities and goals, and 3) The teacher's role. The themes are further elaborated in the below section. In addition, the teachers' feedback to the prototypes is presented as well as the solutions suggested by the teachers who participated in the workshop.

5   Findings

The teachers were all interested in using VR and AR to support their teaching. While some of the teachers had no prior experience with VR and only a few had been using it for teaching, they often introduce and use new technology as part of their teaching. They were therefore open towards using VR as part of the teaching.

5.1   Students and Motivation

In general, teachers noted that there was a need for alternatives to regular classroom teaching, as many of their students were challenged in terms of concentrating and under-standing the material from textbooks. One of the teachers already had good experiences with VR and explained the following:

> *I have experienced students who are scared of technology who distance themselves from it, but I have experienced more students who are turned on academically. I support a curios behavior that makes them achieve a greater learning benefit than previous students have done in my teaching when I have run traditional class- and group work.*

While the teachers see a positive effect on most students, there are still some who find VR transgressive or experience cyber sickness. They also emphasize that it is not the same approach that works with all students. This is in line with the findings in the existing literature [15, 19]. VR should, therefore, not be a standalone option to teaching, but be used as a supplement and be used in specific situations for specific learning goals. This way the different teachers can meet the different traits and characteristics of the students and use VR for students that are motivated to learn from VR as suggested in [19].

5.2 Learning Activities and Goals

One of the situations where the teachers see a potential in VR is for their simulation of diseases. Currently, they use video and pictures or use roleplays where the students or teachers act in the role of an ill citizen. In these situations, they lack a tool to show different diseases and the condition of the citizen which can be used to learn students how to provide the correct care and treatment. There is a general need to be able to visualize different parts of the body both the healthy body and when suffering from a specific disease.

The students often work with case-studies where the teachers present a case about an ill person, but the teachers are lacking visualization tools to support the story. It is easy to read vital signs, measure the weight of a citizen but they must also be able to make observations about the condition of a citizen and make conclusions about a potential treatment in their professional job. It is, however, hard to create realistic and meaningful learning activities that facilitate this type of learning. As expressed by one of the teachers:

We document and measure, but it is important that they can also make observations – see, feel, listen and touch. That is something they should practice more... How is the breathing? it is not just about numbers.

They also talk about being able to have a more holistic understanding of specific conditions and how it affects the citizens e.g. what happens to citizens when they only have 40% lung capacity left due to COPD or what does a person look like with low body mass index (BMI).

Furthermore, the teachers underline the importance of including the affective part of learning, i.e. about understanding what a citizen with COPD goes through, so the student can react in an appropriate way. One of the teachers mention the mental and emotional part this way:

The psychological part in connection with having COPD is important and if you can connect that process, with what he looks like, what does he say... If you could get the emotional part involved, it would make really good sense.

Today, the teachers use videos, along other tools such as a COPD simulation suit or exercises where the students experience reduced lung capacity by breathing through a straw. However, the teachers suggested a VR scenario where sound and the visuals of a citizen with COPD are combined, where the citizen's condition is either improved or worsened by the student's actions. The teachers want a safe environment where students can test different things without fearing the consequences.

5.3 The Teacher's Role

Teachers see themselves as facilitators who must support the students in their learning activities. As the teachers identify poor reflective skills in the students, they often have to assist them with asking questions that facilitate discussions about the new content and how it relates to their future role as caregivers.

Several of the students have difficulty relating to things like science and at times almost experience learning blockages. Teachers therefore feel that they must think alternatively and creatively when they convey part of the curriculum. This is elaborated in the following quote:

> Some students have some unfortunate experiences from their primary school. If I start talking about the periodic table, then I lose half of the students. It is better to start with a different approach and slowly work around it. I often hear from the students that they prefer science when taught at the caregiver education. It is not because I lower the requirements, but because I convey it with a health professional angle.

Instead of learning about physics and chemistry by talking about, for example, formulas, the teachers can show how the different physical and chemical reactions take place in the body using VR and thus students gain an understanding of it. This is what is referred to as non-symbolic interaction by [11]. Instead of learning an abstract symbolic system such as math to explain a chemical process the processes can instead be visualized using VR making it more accessible for the students at the VET colleges.

5.4 Feedback to Existing Prototypes

Overall, the teachers believe that the new solutions should motivate students to learn about the body's anatomy and different conditions from the newly developed prototypes. Many students have a hard time imagining how the different parts of the organs are connected from looking at a picture, as one cannot split the different layers and see what it looks like as with the 3D model in VR.

However, the teachers are not entirely satisfied with the level of detail on the models and the information that comes with it. It might be enough for students who had just started the education, but students later in the education will miss details about e.g. COPD. In the later stage of the education the students need to be able to recognize symptoms through observations and subsequently choose the right care approach and the visualizations in VR did not provide enough details to do so.

In addition, the teachers find that the text boxes contain too much text, and it appear as a wall of text, where the students typically want it to be divided in smaller pieces of information. Instead, they suggest that the students receive information continuously when they interact with the models. In general, they want the students to be more active when using the application.

For the three teachers who participated in the workshop and could try the prototype themselves, they quickly learned how to interact with the models and understood quite quickly how to use gestures and move around the menus.

Several of the teachers suggest having an AR version for phone or tablet that can be used for preparation or reflection before and after teaching, as part of their homework. It was also suggested that sound be used for, for example, to support the simulation of COPD, in which case one would be able to hear the breathing.

5.5 Solutions

Each of the participants in the workshop were asked to draw and describe their ideas to a solution on a piece of paper. Subsequently, they were asked to explain what they had drawn and their idea or concept. The three drawings are shown in Fig. 5, 6 and 7. All teachers chose to present a solution around a case description of a person with an illness. The solutions provided an opportunity to see a holistic picture of the citizen of how the disease affects the person. In all cases, emphasis was placed on the student being able to make choices or actions, which they would then receive feedback on through the system. If they made the right decision, they would e.g. see that the citizen got better. The solution should also include options for varying the level of difficulty. For the teacher with the drawing presented in Fig. 5, the starting point was a citizen with COPD. The student would see an avatar of the citizen indicating how he/she was affected by the disease along with vital sign information and a 3D model of the lungs. This to give a holistic view on how the disease affects the citizen and provides opportunities for the students to combine their knowledge from different courses such as anatomy, pharmacology and somatic illness.

Figure 6 represent a solution where three students are active. One in the role of a citizen, one as a caregiver and one as an observer. The student acting in the role of the citizen has been instructed in the role enabling him or her to answer questions about their wellbeing. Using AR technology, the student would look like an ill citizen when looking through the AR glasses. Thereby AR add a visual dimension to the roleplay that are typically used as part of the teaching to represent a job situation as professional caregiver.

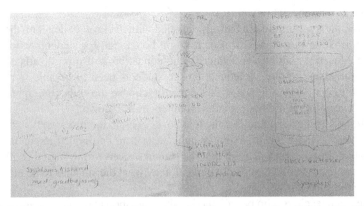

Fig. 5. Solution from the workshop. The student with an avatar of a citizen with COPD. The students can see how the disease effect the avatar, vital signs and an overview of the lungs.

The last solution is in Fig. 7. Again, the student interacts with an avatar of an ill citizen. Based on the interaction with the citizen the student must decide on an action through a multiple-choice menu and again receive immediate feedback on their action.

At some point the actions can lead to the student having a look at the lungs, including aggravations or improvements of the condition, based on the actions or based on the stage of the disease. Furthermore, it was suggested to create two versions of the application one that the students could use with VR glasses and one phone-based solution the students could use to practice at home.

Fig. 6. Solution from the workshop. VR/AR solution that support the student in a roleplay where one student act as the citizen, one as the caregiver and one as observer.

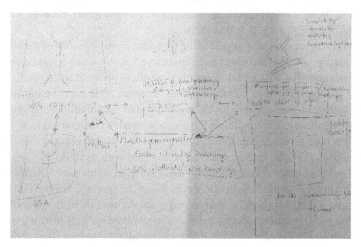

Fig. 7. Solution from the workshop. Again, the students interact with an avatar of a citizen through a multiple-choice menu.

6 Discussion

Several learning goals have been suggested by the teachers where they see a potential for using AR and VR solutions to support the teaching both for cognitive and affective skill training. The overarching learning goal for the students is to be able to observe a citizen's condition and decide on the care or treatment. This encompasses several sub-learning goals. First the students need to know the names of the different body parts and how they function. This include both the healthy body and how a disease progresses and affect the body. Then they must be able to recognize a disease. Understand how they treat and prevent disease, and how diseases affect the citizen both mentally and physically. Finally, the students need to be able to reflect upon these observations and decide on a set of actions to provide the necessary care for the citizen.

The teachers had different suggestions for how VR/AR could be used to support these goals. Firstly, VR/AR can help motive the students to learn by providing the content in a different way, as traditional classroom teaching often works poorly and can lead to learning barriers for many of the students. VR/AR provide options for 3D visualizations that can help students understand the body – how it functions and how diseases progresses. It can be used for homework and repetition training if implemented on a mobile or tablet. It can be used to provide a safe environment for the students to train real life scenarios with interactions and treatment of citizens.

When designing the VR/AR solutions of the future these learning goals should be considered. At the same time, it must be possible to adapt the content and level to the individual student depending on where he/she is in the educational process. It is also important that there is a progression in the solution, so that the degree of difficulty is continuously increased, so the students does not experience the same each time they interact with the system. Teachers put great emphasis on the importance of immediate feedback to guide the students and help them reflect on their experiences in VR/AR and how this could be used in a real-life scenario.

Interestingly the teachers suggested using non-immersive versions of the solution for specific learning goals. This indicates that immersive VR/AR is not always a better solution than non-immersive VR/AR but is highly dependent on the learning goal. Based on the teachers' recommendations and the existing literature on Immersive/non-immersive VR/AR. It seems like immersive VR/AR is more suited for situations that involve both affective and cognitive learning - where the user e.g. needs to understand how symptoms affect a citizen both physical and mentally. Immersive VR/AR is better for simulating social interactions [15], the students feel more engaged with the task [17] and they take the condition of the citizen more seriously [18]. Likewise, immersive VR/AR is recommended for learning how the body works considering the spatial and visual information. Non-immersive VR/AR versions displayed on a screen, mobile or tablet could, on the other hand, be used when only cognitive skills is involved such as learning or remembering the names of the organs or factual knowledge about diseases. This would enable students to use it for preparation and rehearsal as part of their homework.

This hypothesis is based on the assumptions and experiences of the teachers working with the students. Further research including the actual teaching situation in VR and AR – immersive and non-immersive is however, needed. Furthermore, this study did not include the students. Including students in the research would be a next step to study if they achieve the learning outcome that teachers expect.

The research design has been altered due to Covid19. Ideally the initial prototypes should have been designed after gaining the initial feedback from the teachers but instead we choose to use it as a generative tool as part of the interview and workshop.

Due to the limitations of doing the online interview we gained limited feedback in terms of design ideas to the new solution from the teachers who participated in the interview compared to the teachers who participated in the workshop.

7 Conclusion

This study has identified several learning goals that can be supported by VR or AR both cognitive and affective learning goals. The study indicated that the specific learning goals should be considered before deciding on what type of virtual technology and application should be used when designing teaching material for the professional caregiver education. It is important that the solutions offer options for immediate feedback and that the content and difficulty level can be altered to match the educational goals of the student and provide the student with new experiences.

Acknowledgement. This project was partly financed by Videnscenter for Velfærdsteknologi Vestdanmark. The prototypes were developed by Takeawalk VR. Thanks to the teachers from the VET colleges who participated in the study.

References

1. Azuma, R.T.: A survey of augmented reality. Presence Teleoperators Virtual Environ. **6**(4), 355–385 (1997)
2. Sharecare. https://www.sharecare.com/pages/vr. Accessed 26 Oct 2020
3. Yoshida, S., Kihara, K., Takeshita, H., Fujii, Y.: Instructive head-mounted display system: pointing device using a vision-based finger tracking technique applied to surgical education. Videosurgery Miniinvasive Tech. **9**(3), 449 (2014)
4. Pensieri, C., Pennacchini, M.: Overview: virtual reality in medicine. J. Virtual Worlds Res. **7**(1), 1–34 (2014)
5. Green, J., Wyllie, A., Jackson, D.: Virtual worlds: a new frontier for nurse education? Collegian **21**(2), 135–141 (2014)
6. Ma, M., Jain, L.C., Anderson, P. (eds.): Virtual, Augmented Reality and Serious Games for Healthcare 1. ISRL, vol. 68. Springer, Heidelberg (2014). https://doi.org/10.1007/978-3-642-54816-1
7. de Ribaupierre, S., Kapralos, B., Haji, F., Stroulia, E., Dubrowski, A., Eagleson, R.: Healthcare training enhancement through virtual reality and serious games. In: Ma, M., Jain, L.C., Anderson, P. (eds.) Virtual, Augmented Reality and Serious Games for Healthcare 1. ISRL, vol. 68, pp. 9–27. Springer, Heidelberg (2014). https://doi.org/10.1007/978-3-642-54816-1_2
8. Freina, L., Ott, M.: A literature review on immersive virtual reality in education: state of the art and perspectives. In: The International Scientific Conference eLearning and Software for Education, vol. 1, no. 133, p. 10-1007 (2015)
9. Vidensportalen. https://videnscenterportalen.dk/vfv/2019/11/27/viva/. Accessed 26 Oct 2020

10. Khademi, M., Hondori, H.M., Dodakian, L., Cramer S., Lopes, C.V.: Comparing "pick and place" task in spatial augmented reality versus non-immersive virtual reality for rehabilitation setting. In: 2013 35th Annual International Conference of the IEEE Engineering in Medicine and Biology Society (EMBC), Osaka, pp. 4613–4616 (2013). https://doi.org/10.1109/EMBC. 2013.6610575

11. Mikropoulos, T.A., Natsis, A.: Educational virtual environments: a ten-year review of empirical research (1999–2009). Comput. Educ. **56**(3), 769–780 (2011). https://doi.org/10.1016/j. compedu.2010.10.020

12. Winn, W., Windschitl, M., Fruland, R., Lee, Y.: When does immersion in a virtual environment help students construct understanding? In: Proceedings of the International Conference of the Learning Societies, pp. 497–503. Erlbaum, Mahwah (2002)

13. Slater, M.: Immersion and the illusion of presence in virtual reality. Br. J. Psychol. **109**(3), 431 (2018). https://doi.org/10.1111/bjop.12305

14. Pantelidis, V.: Virtual reality in the classroom. Educ. Technol. **33**(4), 23–27 (1993)

15. Jensen, L., Konradsen, F.: A review of the use of virtual reality head-mounted displays in education and training. Educ. Inf. Technol. **23**(4), 1515–1529 (2017). https://doi.org/10.1007/ s10639-017-9676-0

16. Cruz-Neira, C., Sandin, D.J., DeFanti, T.A.: Surround-screen projection-based virtual reality: the design and implementation of the CAVE. In: Proceedings of the 20th Annual Conference on Computer Graphics and Interactive Techniques, pp. 135–142 (1993)

17. Loup, G., Serna, A., Iksal, S., George, S.: Immersion and persistence: improving learners' engagement in authentic learning situations. In: Verbert, K., Sharples, M., Klobučar, T. (eds.) EC-TEL 2016. LNCS, vol. 9891, pp. 410–415. Springer, Cham (2016). https://doi.org/10. 1007/978-3-319-45153-4_35

18. Reiners, T., Wood, L., Gregory, S.: Experimental study on consumer-technology supported authentic immersion in virtual environments for education and vocational training. In: Hegarty, B., McDonald, J., Loke, S. (eds.) Proceedings of the 31st Annual Ascilitite Conference (ascilite 2014): Rhetoric and Reality: Critical Perspectives on Educational Technology, 23–26 November 2014, pp. 171–181. University of Otago, Dunedin (2014)

19. Janßen, D., Tummel, C., Richert, A., Isenhardt, I.: Towards measuring user experience, activation and task performance in immersive virtual learning environments for students. In: Allison, C., Morgado, L., Pirker, J., Beck, D., Richter, J., Gütl, C. (eds.) iLRN 2016. CCIS, vol. 621, pp. 45–58. Springer, Cham (2016). https://doi.org/10.1007/978-3-319-41769-1_4

20. Braun, V., Clarke, V.: Using thematic analysis in psychology. Qual. Res. Psychol. **3**(2), 77–101 (2006)

Designing for Innovation

Towards the Development of AI Based Generative Design Tools and Applications

Juan Carlos Chacón[✉], Hisa Martínez Nimi, Bastian Kloss, and Ono Kenta

Graduate School of Engineering, Chiba University, Chiba, Japan
juancarlos@chiba-u.jp

Abstract. In recent years, several projects that take advantage of Artificial Intelligence as a design tool have arisen. However, most designers lack the technical knowledge necessary to profit from Artificial Intelligence in their design process fully. Through the development of GANSta, a tool with a graphical user interface that facilities the design and training of Generative Adversarial Networks. And the use and application of such a tool in different stages of the design process. By engaging in both iconographic branding element design and typographic font design projects. Participants of the Gesign lab initiative of Chiba University's System Planning Laboratory, explore the current and future opportunities that Generative Adversarial Networks present for their particular design process. Proving that previous knowledge in programming or machine learning is not necessary for designers to take advantage of the benefits that this technology presents from a generative design perspective.

Keywords: Generative adversarial networks · Generative design · Design tools

1 Introduction

Technology has always played a relevant role in the development of design, both at an academic and production level. In recent times, projects like Deepwear [1], which utilizes generative adversarial networks (GANs) to generate images of clothes that a designer can use as the basis for constructing clothing patterns, and the ChAIr Project [2], which uses a similar approach to generate chair shapes that designers can reinterpret and build, have taken advantage of artificial intelligence as a tool within the design process. Researchers such as Kevin German et al. [3] have also proved that AI, particularly GANs, can be used by designers in different stages of the design process.

However, most designers lack the technical skills or knowledge to fully understand the technology behind these projects, often perceiving it as magic [4]. These knowledge and skill gaps represent a relevant barrier that prevents designers, both professionals and those in training, from fully exploiting the potential of these types of AI tools. For that reason, Hughes states that developing design skills and knowledge of transforming technologies is integral for young people to not only understand but also design the future world [4].

E. I. Brooks et al. (Eds.): DLI 2020, LNICST 366, pp. 63–73, 2021.
https://doi.org/10.1007/978-3-030-78448-5_5

In addition, the lack of academic programs and activities that incorporate AI and machine learning technology into the design process has created a need for new learning approaches as well as the development of tools that facilitate the learning and application of AI-based approaches to design.

Thus, within the System Planning Laboratory of the Design Department of Chiba University, the Gesign Lab initiative emerged, involving a small team of academics and students who, through work sessions and small research project workshops, sought to design and develop methods and tools that enabled the use and learning of artificial intelligence concepts, particularly in relation to different design disciplines.

The following research outlines the process and results of the first stage of the project, which ended with the development of GAN Station (GANSta) [5], an internal tool that assists in the design and training of GANs in a graphical environment. This tool resulted from the process of developing two projects, the first focused on the design and generation of logo based graphical symbols through the use of generative adversarial networks, and a second focused on the design of a font character set based on the same approach.

2   Methodology

Project participants were divided into three groups based on their skills and interests: one team focused on tool design and development, and two teams focused on developing projects based on the resulting tools. The latter two teams selected different design outputs based on the personal interests of the participants, similar to a passion-based learning approach [6].

For the development of both projects, different GAN architectures, such as DCGAN [7] and CycleGAN [8], were first explored. After the preliminary results were obtained, StarGAN [9] was selected as the base for the three projects. This decision was made after training a network with a small dataset of 100 black and white images from two different domains, graphically evaluating the results of the interpolation between the domains, and visually and verbally evaluating the results as positive and interesting.

Once StarGAN was selected, a Python-based tool with a text command interface was developed to facilitate training and to test the models generated by the team members who lacked programming skills. This tool took a database folder with a set of images as an input and a simple training and test set of commands. With this first version of the tool available, the teams proceeded to develop the logo based graphical symbol project and the font character design project simultaneously.

2.1   GAN Based Logo Design Project

The objective of this project was to create a new brand logo based on Japanese family crests or Kamons for a Japanese tea shop. Team members proposed using GANs in order to help in the exploration and conceptualization processes and generated several proposals through a series of automated additions, alterations, and combinations of elements, mixing different domains. This approach allowed the team members to generate multiple proposals that, regardless of the selected domain or type of element, could be converted and synthesised to belong to the universe of the current Japanese family crest. During the exploration process, two different experiments were undertaken with different selections of datasets to assess the GAN's creative power. The first exploration process generated new graphic symbols mixing two domains so that the graphic symbol could include elements from two different categories.

The main objective of this first experiment was to generate a new graphic symbol of a teapot that included the characteristics of Japanese Kamons. The dataset was made of 200 black and white graphic symbols of 256×256 pixels that belonged to two different domains: Japanese Kamons and Teapots. As stated before, a StarGAN model was designed and used through the proposed command-based tool. This approach allowed the users to generate images from multiple domains and translate or apply a certain aspect or style from one image to another [10] (see Fig. 1).

In order to achieve this outcome, the team needed to train the generative model (G) to translate an input image x to an output image y conditioned by the target domain. In this first exploration, the total number of parameters for the generator was 8421120. For the discrimination, the total number of parameters was 44786624. The dataset was evaluated with Google Cloud's Vision API, which is called Vision AI [11]. Using this tool for image recognition, the team analysed how 100 images of Kamons and 100 images of teapots were labelled in comparison to the images generated by the proposed model.

The results of the Kamon dataset image analysis were the following: 67% of the images were labelled 'Symbol', 82% were labelled 'Logo', 29% were labelled 'Emblem', and only 5% were labelled 'Crest'. In the case of the teapot dataset, none of the images were labelled 'Symbol', 26% of the images were labelled 'Logo', 98% were labelled 'Teapot', 67% were labelled 'Illustration', and 91% were labelled 'Tableware'.

Finally, after training the discriminator and generator with the small dataset of 200 images, the trained model was tested. The evaluation of the results was performed by a designer through visual inspection, and a set of 100 images were selected to evaluate using Google's Vision API. During the image analysis, 9% of the images were labelled 'Symbol', 25% were labelled 'Logo', 0% were labelled 'Emblem', 69% were labelled 'Illustration', 37% were labelled 'Teapot', and 27% were labelled 'Tableware'. The score of the results was lower on all the labels except for 'Illustration'. Even though the second most recognized label was 'Teapot', none of the images generated by the proposed model were recognized as 'Emblem'. However, despite the automated scores, the designer conducted a visual review and considered some of the results to be interesting visual proposals, concepts, and sketches for the creation of the logo.

For the second exploration, the main objective was to generate a new Kemon incorporating different pre-selected elements and to improve the quality of the images that resulted from the training process. For this goal, a dataset comprised of 640 black and white graphic symbols (256×256 pixels) belonging to 13 different domains was used [12]. The Japanese Kamons were divided into 6 different domain sets due to their similarities in shape and complexity. In addition, 7 domain sets were divided by their visual elements, such as teapots, cups, cakes, trees, leaves, and houses. Those domains were pre-selected by the team members due to their association with the brand personality

Fig. 1. Samples of graphic symbols resulted from the first exploration.

and brand values. The same StarGAN model based on the command line tool was also used, since it was possible to employ multi-domains while generating the images [14].

The GAN was trained using all of the symbols from the 13 domains, with the total number of parameters for the generator amounting to 8452480 and the total number of parameters for the discriminator amounting to 45114304. The evaluation process was done in three stages to analyse the obtained images. For the first stage, the results were evaluated by visual inspection, and 50 different samples were selected (see Fig. 2). Then, for the second stage of evaluation, the results were analysed by the Google Cloud's API, which is called Vision AI [11]. When compared to the first exploration, the second exploration showed a significant increase in label detection, with 32% of the images labelled 'Symbols', 56% labelled 'Logos', 8% labelled 'Emblems', and 82% labelled 'Illustrations'.

Fig. 2. Samples of graphic symbols resulted from the second exploration.

When compared to the results obtained during the first exploration, a significant increase in label detection can be observed. However, a third process was proposed for the evaluation of the results. For this last evaluation phase, a Kamon recognition model was generated to evaluate more accurately if the resulting images belonged to the Kamon category. In this model, 100 images from the Kamon dataset were annotated in the YOLO format and then used for the training of a custom YOLOv5 object recognition model with Kamon as a single category [13].

The resulting model, with a mAP@0.5 of 1.0, was then used to evaluate the 50 results previously selected. According to the results of the Kamon recognition model, 78% of the images were recognised as Kamons with an average confidence rate of 0.7533 over 1.0. This evaluation process contributed to validating whether the images generated by the proposed model could be considered to be symbols, logos, or emblems and specifically if they belonged to the category of Japanese family crests. However, a qualitative evaluation of graphic symbols is also an essential and determining factor in graphic identity design.

In conclusion, the selection of dataset images, the number of domains, and the quality of those images have a direct impact on the quality of the results and are important factors to take into consideration for possible future work. This approach showed that machine learning (and specifically GANs) can be a tool for the branding design process and that graphic design processes could benefit from using GANs to automate and improve their results.

2.2 GAN Based Font Design Project

For the second project in the program, participants envisioned a tool capable of creating a character set with one character as the input. This tool could help a type designer to improve the font creation process by generating a whole character set after only the first character is designed. Similar to the logo project, a Stargate that takes images and manipulates them to fit certain characteristics of different domains, including emotional features on the input image, was designed and used [9, 10, 14].

Using different domains made it possible to create a dimension for every character in the set. Thus, it was possible to take the characteristics of the input and adapt the shape of its different dimensions. To prepare the dataset, the team collected a set of font files. For this project, a master folder of Google fonts was used [15]. The font files were divided into different folders by their category and style.

Aiming for a legible and usable font as the final output of the project, the team decided to only use sans serif fonts in regular style as part of the dataset in order to keep it as simple as possible. All of the glyphs in the font dataset were extracted as PNGs with a size of 256×256 pixels [16]. The dataset was then loaded into the GAN, and the training started. After 1,000,000 training iterations, the characters could be identified in every line produced in a resulting image generated by the GAN. A total of 5,996,000 iterations were trained. Training took 26 days, 10 h, and 27 min until the team decided to interrupt the process for a visual evaluation of the results. Figure 3 shows the results of the 5,400th module.

! " # ? % 3 , ( ) * + , - . /
0 1 2 3 4 5 6 7 8 9 : ; < = > ?
◎ A B C D E F G H I J K L M N O
P Q R S T U V W X Y Z [\ ] ^ _
' a b c d e f g h i j k l m n o
p q r s t u v w x y z { | } ~

Fig. 3. Character samples of the 5,400th training module.

The input that was used to generate these characters was a basic round dot. Figure 4 shows the outputs from the same module with different input variations. The first one shows the result of using a very basic shape, in this case a dot, and the second sheet shows the characters that were generated with a minor äún, äù as the input.

Fig. 4. First output samples generated by the model.

One important finding from this project was that using simpler shapes as the input generated visibly cleaner elements and using more complex shapes caused glitches that resulted in less recognizable characters. For the evaluation, the team created 12 characters of differing complexity to test. The characters were three different versions of a dot, a number one, an exclamation mark, and a minor n. After visual inspection, the team could visually identify that the complexity of the input had a direct influence on the quality of the generated results. The dots, the simplest elements used, had the highest quality outputs.

As the final result, a set of 96 characters corresponding to the basic Roman numerals and letters were generated as a set of PNG files that were then vectorised as a font file, as shown in Fig. 5. Several noise adjustments and flaws, particularly in the baseline and kerning, remained in the final font file as a result of the image generation and vectorising processes. However, as a proof of concept, the team considered the project to be a successful exploration.

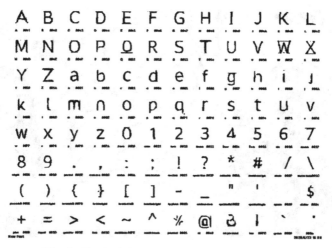

Fig. 5. Final 96 characters generated by the model.

2.3 GAN Station Project

Once the command-based tool was established as the base for the iconographic and typographic projects, the training and evaluation processes for the generated models needed to be improved. To this end, using the code and instructions originally implemented in the command-based tool as a base, a minimum viable prototype of a tool with a graphical interface that would make the process easier and more accessible was developed.

Thus, the GAN Station project, GANSta, emerged, a name voted upon by the participating team members at the time. A node-based interface was adopted due to the familiarity of the team members with this type of interface in tools such as TouchDesigner [17] and Grasshopper [18], as well as for the ease with which these interfaces presented their programming logic, specifically through the use of blocks, which could facilitate the learning of the concepts behind the development of a GAN for future users or participants who lacked prior knowledge on the subject (see Fig. 6).

In the development of GANSta, Anaconda [19] was used with Python 3.7 as the environment, and PyQt [20] was used as the framework for interface building. GANSta's final design consisted of a user interface with the main window divided into a work area and a side menu (see Fig. 7).

Within the work area, users can place, drag, and drop the nodes necessary for the construction of their GAN from the left side window. These nodes consist of the following:

- Dataset node. Its function is to select and verify the path in which the dataset to be used for training is located. This dataset consists of a series of images divided into two folders: one named 'Training' for the images to be used during the training process and one named 'Test' for the images to be used during the test process. Inside the training folder, the folders with the names for each category or domain to be considered must contain their respective images. This node output connects to the Image node.

Fig. 6. Student using the GANstation tool at System Planning Lab. Chiba University, Japan.

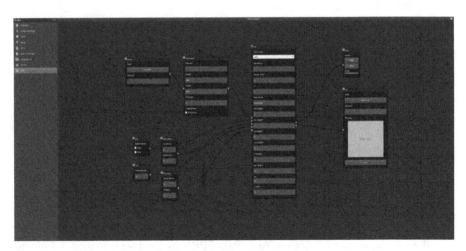

Fig. 7. GANstation user interface.

- Image node. Its function is to define the parameters, such as width and height, of the types of images to be used during the training process and to allow the user to increase those parameters during training so as to distort or randomly rotate the images to avoid problems derived from the lack of variety in the dataset. This node output connects with the GAN node.
- Mode node. Its function is to allow users to select whether they want to train the model or test its results. This node output connects with the GAN node.
- GPU node. This node is optional. Its function is to allow users to select whether or not they want to use GPUs for the training process. It also shows the number of GPUs available on the user's machine. This node output connects with the GAN node.

- Batch node. Its function is to determine the number of batches and epochs needed for the model training. This node output connects with the GAN node.
- GAN node. Its function is to determine the parameters of both the generator and the discriminator, as well as to verify that all of the parameters of the previous nodes are valid or correct. This node output connects with the Output and Action nodes.
- Output node. Its function is to define the path and format of the training results, as well as to provide an internal window to show a preview of the results of the training or testing processes. This node does not connect to the other nodes.
- Action node. Its function is to start the training and testing of the model based on the option selected by the user, as well as to show the percentage of the process completed. This node does not connect to the other nodes.

To use the tool, the user must prepare their respective image-based dataset, as indicated in the dataset node description, and then design its GAN by dragging and dropping the desired nodes into the action window and modifying the parameters of the node as desired. Once the first version of the graphical tool was developed, it was tried and tested by project team members using their existing approach and generated datasets, in order to evaluate the necessary changes in the interface to facilitate its use. As a result of the requests made by the project participants, some changes were made, resulting in a second version of the graphical tool. Among these changes, the Output node was removed, and the functions of both the Output and Action nodes were transferred to a second side menu.

3 Discussions and Conclusion

Although there are other graphical-based tools for programming and interacting with artificial intelligence models [21, 22], the main difference between GANSta and other tools for the development of machine learning models radiates in the fact that GANSta was developed based on the necessities identified during the development of two design-led projects as an internal tool for designers. Seeking to facilitate not only the understanding of the concepts behind the development and operation of the GAN but also facilitating the learning experience, by allowing designers without previous knowledge of programming or machine learning to develop and train their own GAN based models through interactive exploration. Encouraging incorporating the designed tool within their design process.

By designing new artificial intelligence tools to generate iconographic elements and typographic characters for logos and by providing a prepared environment with the necessary instructions to design and train generative adversarial networks (GANs), the participants in the Gesign Lab initiative have proved that both professional designers and design students without prior knowledge of programming or artificial intelligence can identify uses and applications for these tools both in the design process and in the execution of final pieces. However, there are still many limitations beyond the technical requirements of this type of AI tool that may hinder its widespread adoption, such as the cost of the computer equipment, the number of graphic processing units required to execute the training process of the GANs, and the time that the training process takes.

This lengthy process may make a tool such as the one developed for this study unattractive for activities that go beyond experiments or explorations of how design processes could change in the near future. Nevertheless, the fact that designers such as the participants in the Gesign Lab initiative could successfully develop this type of tool shows that it is only a matter of time until artificial intelligence is employed more widely in the design process.

References

1. Kato, N., Muramatsu, N., Osone, H., Ochiai, Y., Sato, D.: DeepWear: a case study of collaborative design between human and artificial intelligence. In: Proceedings of the 12th International Conference on Tangible, Embedded, and Embodied Interaction vol. 2018, January, pp. 529–536 (2018)
2. Schmitt, P., Weiß, S.; The Chair project: a case-study for using generative machin learning as automatism. In: 32nd Conference on Neural Information Processing Systems (NIPS) (2018)
3. German, K., Limm, M., Wölfel, M., Helmerdig, S.: Towards artificial intelligence serving as an inspiring co-creation partner. EAI Endorsed Trans. Creat. Technol. **6**, 162609 (2019)
4. Hughes, J., Robb, J., Lam, M.: Making future-ready students with design and the internet of things. EAI Endorsed Trans. Creat. Technol. **6**, 163096 (2020)
5. Chacón, J.: GAN Station. A node based generative adversarial network tool. (Version 0.1). Zenodo (2020). https://doi.org/10.5281/zenodo.3929374
6. Brown, J.S., Adler, R.P.: Minds on fire: open education, the long tail, and learning 2.0. Educ. Rev. **43**(1), 16–32 (2008)
7. Metz, A.R.L., Chintala, S.: Unsupervised Representation Learning with Deep Convolutional Generative Adversarial Networks. ICLR (2016)
8. Zhu, J.Y., Park, T., Isola, P., Efros, A.A.: Unpaired image-to-image translation using cycle-consistent adversarial networks. In: Proceedings of the IEEE International Conference on Computer Vision, vol. 201, pp. 2242–2251 (2017)
9. Choi, Y., Choi, M., Kim, M., Ha, J.-W., Kim, S., Choo, J.: StarGAN: unified generative adversarial networks for multi-domain image-to-image translation. In: Proceedings of the IEEE Conference on Computer Vision and Pattern Recognition, pp. 8789–8797 (2018)
10. Banerjee, A., Kollias, D.: Emotion Generation and Recognition: A StarGAN Approach, Imperial College London (2019)
11. Google Vision API. https://cloud.google.com/vision. Accessed 23 June 2020
12. Chacon, J., Martinez Nimi, H.: Kamon Dataset and StarGAN Model Version 1.0 Data Set, Zenodo (2020)
13. Chacon, J., Martinez Nimi, H.: yolokamon v1.0 A Yolov5 Japanese Kamon detection model (2020)
14. Choi, Y., Uh, Y., Yoo, J., Ha, J.-W.: StarGAN v2: Diverse Image Synthesis for Multiple Domains (2019)
15. The Google Fonts Team: Google Fonts Files. https://github.com/google/fonts. Accessed 23 June 2020
16. Chacon, J, Kloss, B.: San Serif Font Dataset and StarGAN Model Version 1.0 Data Set, Zenodo (2020)
17. TouchDesigner [Computer software]. https://derivative.ca (2020)
18. Grasshopper [Computer software]. http://www.grasshopper3d.com (2020)
19. Anaconda Software Distribution: Computer software. (Vers. 2–2.4.0). https://anaconda.com (2016)

20. PyQT: Cross-platform software development for embedded & desktop. https://www.qt.io/. Accessed 17July 2020
21. RunwayML [Computer software]. https://runwayml.com (2020)
22. ML5 [Computer software]. https://ml5js.org (2020)

A Model Approach for an Automatic Clothing Combination System for Blind People

Daniel Rocha[1](✉), Vítor Carvalho[1,2], Filomena Soares[1], and Eva Oliveira[2]

[1] Algoritmi R&D, University of Minho, Guimarães, Portugal
id8057@alunos.uminho.pt, fsoares@dei.uminho.pt
[2] 2Ai Lab, School of Technology, IPCA, Barcelos, Portugal
{vcarvalho,eoliveira}@ipca.pt

Abstract. To dress adequately may be a necessary condition in social interaction. The way we dress may have an impact in the way people see us. Recognizing and matching clothes in order to dress properly can be a hard and daily stressful task for blind people. How do they recognize and identify the garments attributes to perform an outfit without help? In order to overcome this stressful situation, we present a project to help blind people in the identification and selection of garments.

Keywords: Convolutional neural networks · Blind people · Clothing recognition

1 Introduction

More and more, aesthetics is present in our lives. The way we dress may have a large impact on our daily lives and may be critical to our well-being.

In fact, our appearance has an impact in the way people see us and the perception they create of us can have a huge influence on the workplace. Some particular events imply dress code and special dedication is devoted to select and combine garments, taking care of all the details from the colours, to the textures, fabrics, accessories, shoes, and scarves. The way we dress is sometimes considered as a business card. This task may be particularly difficult for visually impaired persons, especially if they do it by themselves.

Blind people are dependent on relatives (family, friends) to buy clothes, to detect dirt or to choose the colour of the clothes they want to wear. These are not difficult tasks for those who see, but for those who have little or no vision they become extremely difficult.

For blind people recognizing and choosing an outfit, in order to dress properly, can be a difficult and daily stressful task. Following the previous work [1, 2], this paper is focused in the attempt to recognizing the attributes of the garments to perform later an automatic combination clothing system for blind people.

As mentioned in [2], extracting the clothes features is essential to achieve the goal of automatic combinations. In this sense, in this paper we present the starting point of extracting and classifying the clothing type from an image.

E. I. Brooks et al. (Eds.): DLI 2020, LNICST 366, pp. 74–85, 2021.
https://doi.org/10.1007/978-3-030-78448-5_6

This paper in divided in five sections. Section 2 describes the related work; in Sect. 3 we present the project overview; Sect. 4 explains the methodology, Sect. 5 describes the experiments and finally, Sect. 6 concludes with the final remarks.

2 Related Work

Some fashion apps have been developed as STYLEBOOK, an application to manage the clothes, create outfits, and plan what to wear [3]. ShopStyle [4] allows to plan purchases based on the user's favourite stores, searching items across the web. Another application is Tailor [5] that is a closet that learns the user's preferences and choices and suggests the combinations to wear.

Electronic devices such as Colorino has come to fill the difficulties of the blind in the distinction of colours for the most varied tasks, since it helps in the choice of clothes3, the washing procedure and the colour combination [6]. Another device is the ColorTest 2000, which is a device similar to Colorino that identifies colours and reads the date and time and detects if a light is switched on or off [7].

Yamaguchi et al. [8, 9] propose an approach to clothe analysis. The analysis approach consists in retrieving similar images from a predefined database. The proposed clothing analysis solution is able to classify 53 different categories of fashion photo clothes. The method is able to separate segments in each piece of clothing.

Wazarkar and Keshavamurthy in [10] propose the classification of fashion images, incorporating the concepts of linear convolution and corresponding points using local characteristics. Another study proposes a system for the recognition of patterns in clothes and the dominant colour in each image. The system is a finger-based camera that allows users to query clothing colours and patterns by touch [14].

Yang and Yu [11] propose a real-time clothing recognition method in surveillance settings recognizing eight clothing categories using Histogram of Gradients (HOG) with linear Support Vector Machine (SVM) classifiers.

Yuan, Tian, and Arditi [12] developed a prototype based on computer vision to combine a pair of images of two clothes for pattern and colour. The proposed method for pattern detection achieved 85% accuracy, being robust for clothes with complex texture patterns, various colours and variations in rotation and lighting. For pattern matching, the main errors occurred with images with very similar texture patterns. Regarding colour matching in 100 pairs of images, only one pair did not match correctly due to background distraction. In order to deal with complex texture patterns and lighting changes, they combined techniques using Radon transform, wavelet features, and co-occurrence matrix for pattern matching. The results of the assessment of clothing datasets demonstrate that the method is robust and accurate for clothing with complex patterns and various colours. The corresponding outputs are provided to the user in audio.

Yang, Yuan, and Tian [13] proposes a system for recognizing clothing patterns. This system can identify 11 colours of clothing and recognize 4 categories of clothing patterns. A prototype was developed, based on a camera incorporated in the glasses, a microphone, a computer, and a Bluetooth headset for describing clothing patterns and colours.

The deep learning models dedicated to fashion models prove the importance of neuronal networks in this area. As examples of these works, it is important to refer [14, 15] where the fashion network is based on the VGG-16 architecture. Simonyan and Zisserman [16] introduced a VGG network where they evaluated very deep convolutional neuronal networks for large scale image classification concluding that the depth is beneficial for classification accuracy.

In [17] it is presented a prototype system of clothes detection and classification based on convolutional neural networks. They exceeding F-Score of 0.9 in clothes detection accuracy for five major clothes types.

The work in [18] proposes a method to detect fashion items in a given image using deep convolutional neural networks, achieving 86,7% the average in recall.

Although there are already some studies dedicated to fashion models, there is not yet a solution focused in the development of an automatic system for combining and identifying the garments for blind people.

Considering the research previously carried out, there are tools to directly or indirectly identify colours, patterns, garments but there is not yet available an automatic system for combining and identifying clothing items for the blind. There is where this project stands. The objective is to develop a physical prototype capable of storing garments that identifies the clothes and their wear, dirt, colours and patterns, having the ability to independently suggest clothing combinations to the user.

3 Project Overview

This project follows the work previously carried out, where a combination clothing system for the blind people was created, based on NFC technology (Near Field Communication), placing an identification tag on the garment. With the help of a web application, the blind was able to identify the characteristics of the clothes by reading the label, in addition to managing his/her clothes and combinations[19–21].

The aim now is to introduce artificial intelligence algorithms to suggest combinations and extract characteristics from garments. In order to address this issue of the combination of clothing for the blind people, the development of an autonomous system with artificial intelligence is the basis for the solution to this problem.

As there is still no support system for the blind people, developing a mechatronic prototype system with artificial intelligence will help providing independence and consequent well-being in the identification and combination of clothes, contributing to fill the gap of a technological lack in terms of aesthetics and image of a blind person.

The objective of this work is to develop a prototype that recognizes and takes into account the following elementary requisites:

- Type of clothes;
- The season of the year and weather;
- Suggesting combinations of clothing pieces;
- Identification of the clothe pattern;
- Presence of stains;
- Modifications in the state of the garment.

In pursuit of the objectives, the research question of the overall project arises:

- How can a mechatronic device with artificial intelligence make the inspection, identification, combination and management of clothing for a blind person?

It is worth mention that this work has the collaboration of the Association of the Blind and Amblyopes of Portugal (ACAPO). The main objective of the partnership is to design, enhance and validate all the work that is being developed.

4 Methodology

The literature review allowed to identify the research opportunity, as the existing technologies and investigations are few and limited.

In order to achieve the proposed objectives, a deductive approach is made, formulating hypotheses based on a critical analysis of the literature review, since it is based on scientific principles.

The research strategy adopted, the action research is in accordance with the research question and the final objective, in which a prototype must be reached, Fig. 1.

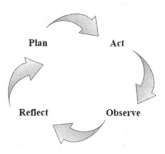

Fig. 1. Action research cycle.

Given that it is intended to develop a technology with practical application in the real context, this strategy is supported by active research, with the participation and collaboration of the target audience.

The process of completing the prototype includes at least two iterations, meaning that two cycles from the beginning to the end are needed to obtain significant improvements. It means each time a cycle is repeated the set of improvements introduced, in order to obtaining the objectives initially stated, tends to stabilize. Qualitative and quantitative data are used to assess the usability and performance of the prototype.

In the development of the tool for image acquisition at an early stage, a mobile device will be used due to accessibility and usability.

The output from an artificial intelligence algorithm will feed the entire system in order to create a database. To obtain the combinations we need to characterize and recognize the characteristics of the garments, as well as, to have a description of the garments contained in the wardrobe. The latter is considered important for the user.

The entire process from acquiring an image to obtaining a combination is illustrated in Fig. 2.

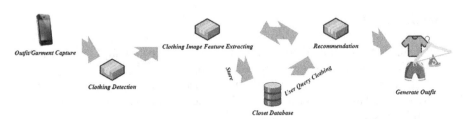

Outfit/Garment Capture   *Clothing Image Feature Extracting*   *Recommendation*

Clothing Detection   *Store*   *User Query Clothing*   *Generate Outfit*

Closet Database

Fig. 2. Workflow for extracting the clothing attributes and recommendation.

The image of the clothing can be obtained through a photograph or an online store, where it is processed through Convolutional Neural network (CNN) to classify and extract the characteristics of the garment. Subsequently, its characteristics will be saved in the database to be used during the combination task. This database represents the virtual wardrobe of the blind user.

The high-level steps associated to the clothing combination are described as:

- Clothing Detection: recognize/segment the garment contained in the photography and assign it to the categories upper, lower, shoes or accessories.
- Clothing Image Feature Extraction: extract features/attributes of the garment as: type, season, pattern.
- Recommendation: suggest a combination for a garment, fulfil an incomplete outfit and suggest a complete outfit.

Neural networks will be used for clothing classification and combination. For the extraction of characteristics, CNNs will be used due to the good results demonstrated in the state of the art.

During a process of combinations, the algorithm will learn the user's preferences based in its frequency of use.

As the target group is blind people and in order to provide good usability practice, a friendly interface will be created, providing mechanisms to assure the proper accessibility by smartphone or tactile screen. A database will be designed to record all data processed and acquired, including the results of statistical analysis. The software will be designed with multiplatform compatibility, in the sense of being later integrated into several operating systems.

Later, a mechatronic system with robotics will be designed in order to integrate all the developed software. This system will make the entire process of choosing, identifying and selecting garments autonomously.

Finally, test trials are essential to assess the proper functioning of the system. Several tests will be considered involving a sufficient number of blind people, as well as worst-case scenarios to test the robustness of the algorithms for real time image processing and data acquisition. This task must act in an iterative approach with the other tasks, in order to allow the optimization of the system, since the collection of quantitative data will be essential until obtaining the final version.

5   Experiments

In this work we decided to build a dataset in order to identify how its construction can influence the results.

The dataset was based on images collected from the internet and distributed in 7 categories, according to the Table 1. In online shops and e-commerce stores the photos have a white background with only one item of garment by each photo, that is exactly the opposite of the pictures in our dataset where most of them have a background and more that one part of the body.

Table 1. Dataset description

Category	Number of samples
Ankle boot	105
Bag	105
Coat	105
Pullover	105
Sneaker	105
T-Shirt	105
Trouser	105

In order to achieve good results, it is essential to take into consideration image processing before introducing the images in neuronal networks. However, we try to approximate our dataset to possible real-life cases that could occur with the user taking a photo of his/her clothe. In addition, image processing requests more resources and time.

Based in the success of the works mentioned in the state of the art we adopted in our approach the convolutional neuronal networks for feature extracting and image classification.

In order to evaluate the data and starting sketch our algorithm we choose two types of CNN applied in ImageNet Large Scale Visual Recognition Challenge [22]. The goal of ImageNet is object detection and image classification where an input image is classified in one of the 1,000 categories.

In this context, we have implemented the VGG16 from scratch in Keras due to building simplicity using 3×3 convolution layers with a softamx classifier after the fully connected layers as show in Fig. 3.VGG means Visual Geometry Group at University of Oxford and the number "16" represents the weight layers in the network.

ConvNet Configuration					
A	A-LRN	B	C	D	E
11 weight layers	11 weight layers	13 weight layers	16 weight layers	16 weight layers	19 weight layers
input (224 × 224 RGB image)					
conv3-64	conv3-64 **LRN**	conv3-64 **conv3-64**	conv3-64 conv3-64	conv3-64 conv3-64	conv3-64 conv3-64
maxpool					
conv3-128	conv3-128	conv3-128 **conv3-128**	conv3-128 conv3-128	conv3-128 conv3-128	conv3-128 conv3-128
maxpool					
conv3-256 conv3-256	conv3-256 conv3-256	conv3-256 conv3-256	conv3-256 conv3-256 **conv1-256**	conv3-256 conv3-256 **conv3-256**	conv3-256 conv3-256 conv3-256 **conv3-256**
maxpool					
conv3-512 conv3-512	conv3-512 conv3-512	conv3-512 conv3-512	conv3-512 conv3-512 **conv1-512**	conv3-512 conv3-512 **conv3-512**	conv3-512 conv3-512 conv3-512 **conv3-512**
maxpool					
conv3-512 conv3-512	conv3-512 conv3-512	conv3-512 conv3-512	conv3-512 conv3-512 **conv1-512**	conv3-512 conv3-512 **conv3-512**	conv3-512 conv3-512 conv3-512 **conv3-512**
maxpool					
FC-4096					
FC-4096					
FC-1000					
soft-max					

Fig. 3. ConvNet configurations [16]

The dataset was split in 80% for training and 20% for testing.

The model achieved 80% accuracy in the training set and 66% in the testing set.

Figure 4 shows the behaviour relative to the accuracy and loss during the training and testing process. The graph shows that the best accuracy was achieved with 30 epocs, while obtaining the lowest loss value in the model.

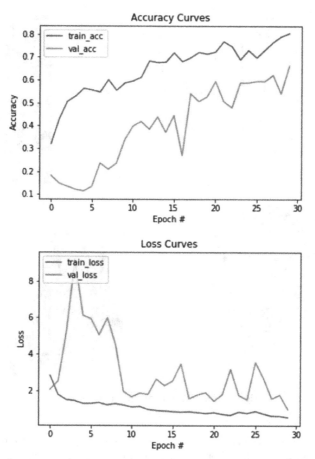

Fig. 4. Loss and accuracy of VGG16 model (training process in blue and testing process in orange).

After the training and validation phases, some images were submitted to the model for classification. Examples of classification can be verified in Fig. 5. Although in Fig. 5 are shown good classification scores, it was also observed some misclassification in particular, in garments with similar shapes.

A recent Google study shows that training from fine tuning is better than training form scratch since it is required less cost in terms of data and time to training [23]. In this sense, a ResNet50 [24] pretrained ImageNet weights was used. It was replaced the top of original architecture, the fully-connected layers and softmax classifier by new ones in order to learn the new classes. In this way only a head part of the network is re-trained.

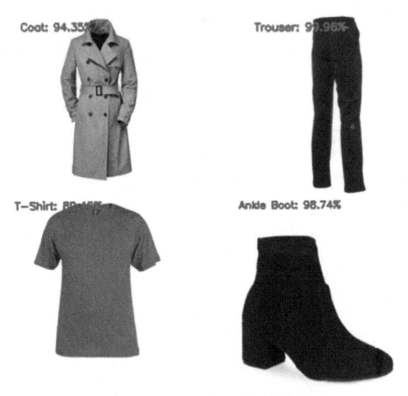

Coat: 94.33%   Trouser: 99.96%

T-Shirt: 88.46%   Ankle Boot: 98.74%

Fig. 5. Example of garment classification.

The fine-tuning of ResNet50 was trained under Google Colab that is a free cloud service hosted by Google. The major advantage of this notebook is that it provides a free GPU [25].

The results obtained with this fine-tuning training are approximated to the previous network where 63% accuracy in training set and 71% in accuracy test set were achieved. Figure 6 shows the behaviour relative to the accuracy and loss during the training and testing process in ResNet50. It is possible to observe that during the build of the model, accuracy and loss in test set have achieved a better result than in the training set.

Comparing the graphs of the training and validation accuracy and loss from both configurations, it is possible to see that the ResNet50 needs more epochs to achieve the same accuracy, Fig. 6. Note that in the VGG16 with 30 epochs it is achieved the best performance possible. Otherwise the last one needs 60 epochs to achieve the same results.

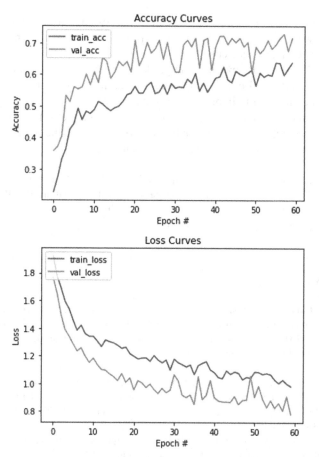

Fig. 6. Accuracy and loss in ResNet50 (training process in blue and testing process in orange).

6   Final Remarks

It is recognized that the aesthetics of clothing is an important issue in human lives.

Blind people may face additional problems in accomplishing this issue, namely in identifying and combining clothing. Some are helped by family or friends, others need a pronounced organizational capacity. This system proposes a new concept to significantly improve the daily life of blind people, allowing the blind to identify and combine clothing pieces.

During this paper we have explored two different convolutional neuronal networks being one of them implemented from scratch and the other one through fine-tuning.

The amount of data on each category is not enough to reach good results. The deep learning is adequate for big amount of data, meaning that it is necessary to gather more data in order to improve the results.

The results obtained by using a simple network designed from scratch and with a fine-tuning network were similar. Moreover, to process the fine tuning in ResNet50 has higher computational and time costs.

In the future, we will experiment large datasets, extracting and identifying new attributes. Some additional image processing, like remove background and body parts will be performed in order to improve the results.

Finally, we intend to implement our algorithm on a web platform or mobile device, in order to start testing interactivity with blind users. After that, the smart wardrobe will also be built to incorporate all the developed algorithms.

Acknowledgments. This work has been supported by COMPETE: POCI-01–0145-FEDER-007043 and FCT – Fundação para a Ciência e Tecnologia within the Project Scope: UID/CEC/00319/2020. The authors would like to express their acknowledgments to Association of the Blind and Amblyopes of Portugal (ACAPO).

References

1. Rocha, D., Carvalho, V., Soares, F., Oliveira, E.: Design of a smart mechatronic system to combine garments for blind people: first insights. In: Garcia, N.M., Pires, I.M., Goleva, R. (eds.) HealthyIoT 2019. LNICSSITE, vol. 314, pp. 52–63. Springer, Cham (2020). https://doi.org/10.1007/978-3-030-42029-1_4
2. Rocha, D., Carvalho, V., Soares, F., Oliveira, E.: Extracting clothing features for blind people using image processing and machine learning techniques: first insights. In: Tavares, J.M.R.S., Natal Jorge, R.M. (eds.) VipIMAGE 2019. VipIMAGE 2019. Lecture Notes in Computational Vision and Biomechanics, vol. 34, pp. 411–418. Springer, Cham (2019). https://doi.org/10.1007/978-3-030-32040-9_42
3. Stylebook Closet App: a closet and wardrobe fashion app for the iPhone, iPad and iPod. http://www.stylebookapp.com/index.html. Accessed 21 June 2019
4. ShopStyle: Search and find the latest in fashion. https://www.shopstyle.com/. Accessed 21 June 2019
5. Tailor : The smart closet. http://www.tailortags.com/. Accessed 21 June 2019
6. Colorino Color Identifier - Light Detector - Assistive Technology at Easter Seals Crossroads. https://www.easterealstech.com/2016/07/05/colorinos-color-identifier-light-detector/. Accessed 19 Feb 2019
7. Colortest Standard - Computer Room Services. https://www.comproom.co.uk/product/colortest-classic/. Accessed 19 Feb 2019
8. Yamaguchi, K., Kiapour, M.H., Ortiz, L.E., Berg, T.L: Parsing clothing in fashion photographs. In: 2012 IEEE Conference on Computer Vision and Pattern Recognition (CVPR), pp. 3570–3577 (2012)
9. Yamaguchi, K., Kiapour, M.H., Ortiz, L.E., Berg, T.L.: Retrieving similar styles to parse clothing. IEEE Trans. Pattern. Anal. Mach. Intell. **37**, 1028–1040 (2015). https://doi.org/10.1109/TPAMI.2014.2353624
10. Wazarkar, S., Keshavamurthy, B.N.: Fashion image classification using matching points with linear convolution. Multimedia Tools Appl. **77**(19), 25941–25958 (2018). https://doi.org/10.1007/s11042-018-5829-4
11. Yang, M., Yu, K.: Real-time clothing recognition in surveillance videos. In: 2011 18th IEEE International Conference on Image Processing (ICIP), pp. 2937–2940 (2011)

12. Yuan, S., Tian, Y., Arditi, A.: Clothing matching for visually impaired persons. Technol. Disabil. **23**, 75–85 (2011). https://doi.org/10.3233/TAD-2011-0313
13. Yang, X., Yuan, S., Tian, Y.: Assistive clothing pattern recognition for visually impaired people. IEEE Trans Hum-Mach. Syst. **44**, 234–243 (2014). https://doi.org/10.1109/THMS.2014.2302814
14. Wang, W., Xu, Y., Shen, J., Zhu, S.-C.: Attentive fashion grammar network for fashion landmark detection and clothing category classification. In: Proceedings of the IEEE Computer Society Conference on Computer Vision and Pattern Recognition, pp. 4271–4280 (2018)
15. Liu, Z., et al.: DeepFashion: powering robust clothes recognition and retrieval with rich annotations. In: 2016 Conference on Computer Vision and Pattern Recognition (CVPR), pp. 1096–1104 (2016)
16. Simonyan, K., Zisserman, A.: Very deep convolutional networks for large-scale image recognition. In: 3rd International Conference on Learning Representations (ICLR 2015) - Conference Track Proceedings (2015)
17. Cychnerski, J., et al.: Clothes detection and classification using convolutional neural networks. In: 2017 22nd IEEE International Conference on Emerging Technologies and Factory Automation (ETFA), pp. 1–8 (2017)
18. Hara, K., Jagadeesh, V., Piramuthu, R.: Fashion apparel detection: the role of deep convolutional neural network and pose-dependent priors. (2014). https://doi.org/10.1109/WACV.2016.7477611
19. Rocha, D., Carvalho, V., Gonçalves, J., Azevedo, F., Oliveira, E.: Development of an automatic combination system of clothing parts for blind people: MyEyes. Sensors Transd. **219**(1), 26–33 (2018)
20. Rocha, D., Carvalho, V., Oliveira, E., Gonçalves, J., Azevedo, F.: MyEyes-automatic combination system of clothing parts to blind people: first insights. In: 2017 IEEE 5th International Conference on Serious Games and Applications for Health (SeGAH), pp 1–5. IEEE (2017)
21. Rocha, D., Carvalho, V., Oliveira, E., Gonçalves, J., Azevedo, F.: MyEyes-automatic combination system of clothing parts to blind people: first insights. In: 2017 IEEE 5th International Conference on Serious Games and Applications (SENSORDEVICES 2017), Italy, pp. 11–14 September 2017
22. ImageNet Large Scale Visual Recognition Competition (ILSVRC). http://www.image-net.org/challenges/LSVRC/. Accessed 13 July 2020
23. Kornblith, S., Shlens, J., Le, Q.V.: Do better ImageNet models transfer better?. In: 2019 IEEE/CVF Conference on Computer Vision and Pattern Recognition (CVPR), pp. 2656–2666 (2018). https://doi.org/10.1109/CVPR.2019.00277
24. He, K., Zhang, X., Ren, S., Sun, J.: Deep residual learning for image recognition, In: 2016 IEEE Conference on Computer Vision and Pattern Recognition (CVPR), pp. 770–778 (2016). https://doi.org/10.1109/CVPR.2016.90
25. Welcome To Colaboratory - Colaboratory. https://colab.research.google.com/notebooks/intro.ipynb. Accessed 14 July 2020

Seasonal Sunlight Chamber: A Lighting Design Concept to Connect Us to the Dynamics of Sunlight and Our Place on Earth

Emma Strebel[✉] and Ellen Kathrine Hansen

Lighting Design, Department of Architecture, Design and Media Technology, Aalborg University, A. C. Meyers Vænge 15, 2450 Copenhagen, Denmark
ekh@create.aau.dk

Abstract. People currently exist mainly indoors, detached from their natural surroundings. During times of rapid growth, globalization and digitalization, it has never been more important to investigate how to reconnect to our natural environment. In this paper we develop a design to investigate how a lighting design concept can act as a tool to understand the geometry of sunlight on Earth and thereby meet human needs to be in touch with the environment. A design is developed by redefining an ancient analogue technology, the sundial. The path of the Sun is translated into a design concept and is demonstrated in a three dimensional time and sight specific prototype. This design concept creates embodied experiences where viewers interact with their ever-changing daylight and surroundings. The aim with this exploratory design is to create a visual tool for learning about complex natural phenomena and understanding our relation to Earth and the Sun. It thereby discusses how a design can put humans in touch with their natural surroundings to satisfy individual biological needs in order to better understand contemporary environmental needs at large.

Keywords: Daylight design · Design research · Natural phenomena · Dynamic sunlight · Dynamic daylight · Sundial · Changing surroundings · Environmental change · Seasonal change · Circadian rhythm · Diurnal rhythm · Lighting design · Exploratory design · Teaching tool

1 Introduction

Exploring the human connection to diurnal changes in daylight is an integral aspect of human evolution and is vital in the development of this design in order to fulfill human needs. Research starts with the impact that the daily and yearly rotations of the earth have on human evolution. The earth rotates at a speed of one rotation around its axis every day and one rotation around the Sun every year. At the root of it, these changes are physical relations between one another. The day, month and year cycles are not tied to time, but they are tied to gravity and laws of motion [1]. In order to survive on Earth, all life, including humans, evolved to adapt to these kinds of external changes: *"For millennia*

© ICST Institute for Computer Sciences, Social Informatics and Telecommunications Engineering 2021
Published by Springer Nature Switzerland AG 2021. All Rights Reserved
E. I. Brooks et al. (Eds.): DLI 2020, LNICST 366, pp. 86–98, 2021.
https://doi.org/10.1007/978-3-030-78448-5_7

before clocks, our only regular way of measuring time had been the alternation of day and night. The rhythm of day followed by night also regulates the lives of plants and animals. Diurnal rhythms are ubiquitous in the natural world. They are essential to life [...]. Living organisms are full of clocks of various kinds - molecular, neuronal, chemical, hormonal - each of them more or less in tune with the others" [1].

As day follows night, our hormones adapt, influencing our neurons, triggering a release of chemicals, resulting in behavior, etc. It is a constant feedback loop. The circadian rhythm exemplifies that the ability to perceive change is at the core of what makes an organism successful at living on Earth. We have developed ways to detect changes that occur in our surroundings. The sense which evolved most with the diurnal cycle of change is the eye's ability to see light. *"Sight like hearing requires a modulated and crafted form of light for meaning. Stabilize images perfectly on the retina and they disappear. This is a fact of sense psychology. We see only change, movement, life"* [2].

Humans have evolved to have a keen sense for earthly changes. However, changes that happen at a slower time-scale, become visually unnoticeable such as a tree budding or ice caps melting. In this paper, these changes are defined as "earthly changes". While earthly changes are not visually noticeable to the human eye, we have nevertheless evolved an acute ability to register them and they are vital to humans' ability to perceive time. Our brain catalogues the gestalt feeling of a space. As Rovelli says, *"The past leaves traces of itself in the present"* [1]. We are full of memories which are traces from the past and inform our present experience of our surroundings. The American researcher and lighting designer William Lam continues this thought when he says, *"Our evaluation of any environment is colored by the memory of prior experience in analogous situations"* [3]. Humans constantly read their lit environment in order to acclimate to a space. Lam writes about lighting for humans' biological needs. Among others, *"Location, with regard to [...] sunlight"* and *"Time, and environmental conditions which relate to our innate biological clocks"* are two major biological needs [3]. Lam references Vernon as he says, *"The type of motivation to which perception is mostly directly related is the necessity of maintaining contact with the environment and adapting behavior to environmental change"* [4]. The circadian rhythm is a prime example of our biological ability to notice slow, earthly changes. By evaluating one's lit surroundings, the circadian rhythm allows humans to prepare the body for changes to come.

While there is a strong biological need to connect to earthly changes in daylight, there is an even greater need for humanity to understand earthly changes in order to sustain environmental equilibrium on Earth. Instant gratification, more than ever before, seduces people's focus, in turn causing us to disregard the experience of earthly changes. As the astronomer Carl Sagan says, *"We are very devoted to the short-term and hardly ever think about the long-term"* [5]. A focus on slower, environmental changes is being overlooked. With this mass cultural switch, people have become more detached from their place on Earth and the rhythm of the day and year. This becomes particularly crucial when conceptualizing the interconnectivity of humans and the planet which is vital in order to understand the environmental impact humans have on Earth. While seasonal climates are changing, ecosystems are thrown off balance [6]. Understanding the interconnectivity between humans and the Earth's ecosystems is the first step toward managing a healthy and sustainable environment.

This paper illustrates how a design concept can build a connection between individuals and their environment. Donella H. Meadows explains how we operate within an interconnected world, full of systems which have built our belief systems, physical relations and natural surroundings [7]. Life as we know it is held together by these ever-changing feedback-loops. While generating substantial change is difficult, Meadows says there are different leverage points to *"change the structure of systems"* [7]. The most impactful instrument of change is to alter a mind-set—a *"paradigm"* [7]. This, she argues, is not quantifiable, it is a belief system, an outlook. For an individual, *"All it takes is a click in the mind [...] a new way of seeing."* [7]. We have all experienced the instant when something clicks into place: comprehension is unlocked and we truly understand something new. These moments often occur during experiences. Experiential learning is a powerful method for learning and *"take[s] place between individuals and the environment."* [8]. While developing a design, the goal is to create an experience that puts individuals in touch with their environment in order to see their surroundings in a new way. The design approach is inspired by the Exploratorium where *"exhibits [...] let you interact directly with real phenomena"* to *"create inquiry-based experiences that transform learning"* in order to encourage viewers to *"confidently ask questions, question answers, and understand the world around them"* [9]. In this project, an exploratory design is used in order to investigate a natural phenomena – sunlight's relationship to earthly changes. By learning through exploratory design, viewers can build an appreciation for the immense system on which we live—Earth.

This paper is based on a MSc thesis of Lighting Design at Aalborg University by the first author [10]. It leads to a question for investigation: How can a lighting design accentuate changing moments of sunlight, thus building an understanding of peoples' place on Earth? Continued research explores ways of using ancient tools that accentuate a changing moment of sunlight in order to develop a contemporary and innovative design that connects people with their place on Earth.

2   Sundial: A Tool to Visualize Change

The Sundial is an ancient analogue technology that uses the predictable relationship between Earth and the Sun to visualize change. It utilizes the straight rays of sunlight in order to mark daily and yearly changes. Blocking sunlight creates a sharp shadow with high contrast between light and shadow. Since sunlight-shadows create sharp edges, one is able to notice the position of the Sun in relation to an object based on the shadow that it casts. In a sundial, the position of the shadow is contextualized in the cycle of the day and year. A mark of sunlight exists in relation to the rotation of Earth around its axis and around the Sun. By using this astronomical tool, we are able to predict where the Sun will be in the sky at any given moment in any place on Earth. By marking the position of sunlight at a given moment, it creates a visual reference to a moment of light in the past and future. It calls attention to the earth's routine of light. While sunlight is not always visible because of weather conditions, the eart's rotation allows for the Suns movement and position to be predictable throughout a space. The predictability of sunlight's movement will be referred to as the "geometry of sunlight". Sundials use the geometry of sunlight to offer a visual and spatial orientation to the time of day and year

and place on Earth. The concept of this ancient technology is incorporated in the design development in order to spatially visualize the daily and yearly cycles.

The sundial is traditionally used to quantify change rather than to create an experience of change. However, the experience of earthly changes, particularly in daylight, is not only a human biological need [3], but is vital in understanding the timescale of contemporary environmental changes. Occasionally, there are moments during the day and year that accentuate poignant experiences of light, such as at sunset. Encounters like these remind us of the unique beauty of a particular moment on Earth. Witnessing a precise moment of light draws attention to the change inherent in daylight. Learning from these experiences of unique moments of light ground us in our ever-changing physical surroundings. Concepts of sundials appear in art and architecture throughout time.

"[Light] has been treated scientifically by physicists, symbolically by religious thinkers, and practically by artists and technicians. Each gives voice to a part of our experience of light. When heard together, all speak of one thing whose nature and meaning has been the object of human attention and veneration for millennia. During the last three centuries, the artistic and religious dimensions of light have been kept severely apart from its scientific study. I feel the time has come to welcome them back, and to craft a fuller image of light than any one discipline can offer" [2].

Artists and architects accentuate the dynamics of daylight in order to connect viewers to their place on Earth. The following artists, architects and historic sites act as inspiration for developing an exploratory design that focuses on creating a space to experience changes in sunlight.

A historic monument in Ireland called Newgrange, is aligned with the mid-winter sunrise. During this time, sunlight enters 18 m through the space, clearly accentuating a magical moment during Ireland's dark winters [11]. James Turrell creates *Skyspaces* to observe the dynamics of daylight [12]. In the stillness of his architectural spaces, the changes in daylight become central. Chris McCaw creates photographs where the Sun burns a hole into a film negative creating a time lapse or map of the Sun's movement throughout the course of a day [13]. The Finnish architect Pallasmaa describes natural light as a vital light source for humans but also as something we often take for granted. He expresses excellent architecture as something that makes us aware of the surroundings by letting us experience light in all its nuances and feel its presence in space [14]. Lastly, the Danish architect Jan Utzon explores this phenomenon in his own house in Mallorca. The house is built to accentuate the dynamics of daylight over the course of the day and year [15].

Referring to phenomenological philosophy, these projects do more than just create a visual experience of the movement of light. Light interacts with the space, creating embodied experiences from which to learn from while being present in the space [16]. They develop atmospheres that change with their viewers.

3   The Exploratory Design

3.1   Design Concept

To investigate how to create a better understanding of earthly changes an exploratory design is developed. The method refers to Schöns reflection-in-action [17] and the Exploratorium [9]. The concept for the exploratory design is to investigate how to use sundial technology to develop a design that meets the human biological need for experiencing and understanding earthly changes in daylight [3]. By experiencing changes in sunlight, viewers observe their connection to their surroundings. This design sets out to investigate natural phenomena and create a visual experience of *"change, movement, life"* [2]. Experiencing these rhythms of light connects people to rhythms of life. As daylight slowly changes, people's emotional and physical states change accordingly. Based on the research presented in the introduction and the artistic and architectural inspirations derived from sundials, a set of criteria are determined for creating the design concept: the design must be created in relation to the geometry of sunlight throughout a space and use sunlight to accentuate daily and seasonal changes.

Using these criteria, a structure is developed to create a space where viewers experience changes in sunlight. The design is built based on its relation to the Sun and exemplifies direct sunlight as it enters into a structure. Sunlight enters into a long rectangular opening on the top of the space and casts a stripe of sunlight down the side and bottom of the structure. Over the course of a day, the stripe of sunlight enters through the opening, and moves down the west wall. It aligns with the floor precisely at midday. After midday the sunlight moves up the east wall. Midday is the only time when light does not hit the walls (see Fig. 1).

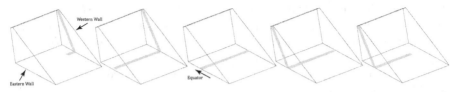

Fig. 1. Figure shows the sunlight as it moves through the space before midday (left), at midday (center) and after midday (right). The opening faces toward the equator (180°S in the Northern Hemisphere and 0°N in the Southern Hemisphere). This image depicts the light moving through the space in the Northern Hemisphere. Figure by Emma Strebel.

Over the course of six months, sunlight moves from the southernmost point to the northernmost point of the floor. Every year, the sunlight moves throughout the space and back again, filling the entire space with light twice per year (see Fig. 2).

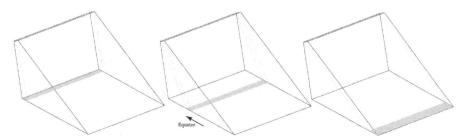

Equator

Fig. 2. Figure shows the sunlight as it moves through the space at midday over the course of 6 months from June solstice to December solstice. Figure by Emma Strebel.

The seasonal geometry of sunlight determines the angles of the north and south walls of the structure. Depending on the latitude of where the piece is installed, the angles of the north and south walls are different. While the size of the opening and the difference between the angles of the north and south walls stay the same with a consistent 47°, the shape of the entire structure changes relative to the ground (see Fig. 3 and 4). In Denmark, for example, the midday sun on June solstice has an altitude of 58° which is the angle of the south wall, and the midday sun on December solstice has an altitude of 11° which is the angle of the north wall (see Fig. 5).

The north and south walls are determined by the angle of the sunlight at midday on June solstice and December solstice. No matter where one is on Earth, the difference between these angles is 47°. The area that is created is a seasonal sunlight chamber which becomes the design (see Fig. 3, 4 and 5).

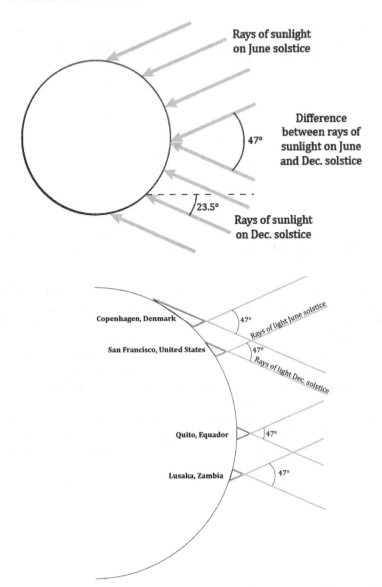

Fig. 3. Figure shows the change in angle of sunlight depending on season. The rays of light on the solstices create the angles of the north and south walls of the structure. Since sunlight hits the earth in parallel lines, the difference in the angle of sunlight hitting the earth between the solstices is always 47°. Figure by Emma Strebel.

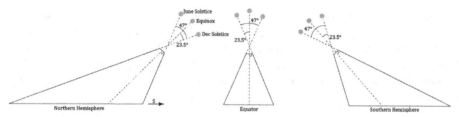

Fig. 4. Figure shows the general shape of the structure depending on whether the design is located north, south or on the equator. Depending on where the structure is located, the opening faces a different direction. The opening is always perpendicular to the angle of sunlight at midday on the equinox. Figure also shows that the north and south walls are angled with 47° difference. Figure by Emma Strebel.

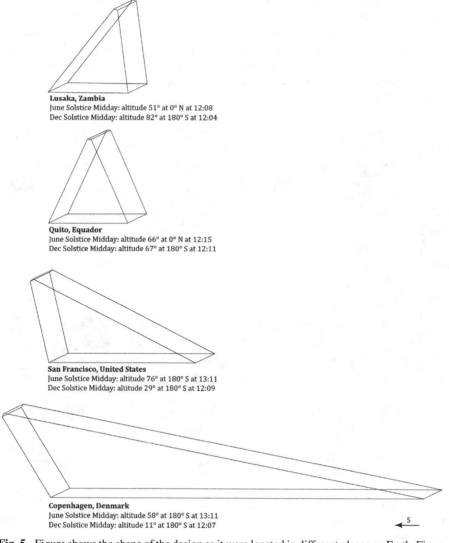

Lusaka, Zambia
June Solstice Midday: altitude 51° at 0° N at 12:08
Dec Solstice Midday: altitude 82° at 180° S at 12:04

Quito, Equador
June Solstice Midday: altitude 66° at 0° N at 12:15
Dec Solstice Midday: altitude 67° at 180° S at 12:11

San Francisco, United States
June Solstice Midday: altitude 76° at 180° S at 13:11
Dec Solstice Midday: altitude 29° at 180° S at 12:09

Copenhagen, Denmark
June Solstice Midday: altitude 58° at 180° S at 13:11
Dec Solstice Midday: altitude 11° at 180° S at 12:07

Fig. 5. Figure shows the shape of the design as it were located in different places on Earth. Figure by Emma Strebel.

The variations in the shape of the structure visualize the acute differences in sunlight based on its location on Earth. Far from the equator, the angle of the sunlight is shallower and thus, the angles of the north and south walls are shallower. Since the structure is built in relation to its location on Earth, each aspect of the design is dependent on that location. Regardless of whether the sun is shining or not, the differences in the design's structural form visualizes the differences in sunlight at different places on Earth.

3.2 Prototype

A prototype is created to examine how the sunlight moves throughout the space. The calculations of the north and south walls are specific to the location of San Francisco. The mockup is made out of diffuse plexiglass so that the structure is not transparent but the direct sunlight is visible from the outside. In this iteration, the prototype hangs from two C-stands so that the opening faces 180° south and the bottom is level to the ground. Once the prototype is set up, the movement of the sunlight is documented through the structure close to June solstice (see Fig. 6) and near December solstice (see Fig. 7). Due to the parallel rays of sunlight, this design is scaleable to any dimension. While this prototype is under a cubic meter, the design would operate the same way as a room viewers could enter into.

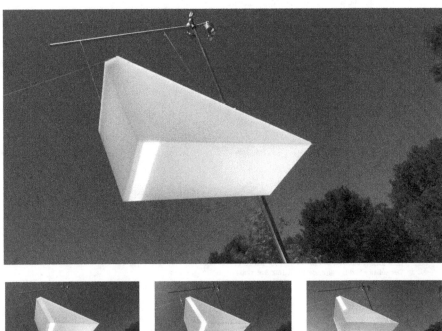

Fig. 6. Photos demonstrate the design prototype using Lighting White 60% plexiglass. The top image shows light entering into the structure precisely at midday near to June solstice. The following three photos show the light as it enters into the structure after midday. Photos taken by Emma Strebel.

Fig. 7. Photo shows light entering into the structure at midday as the season nears December solstice. Photo taken by Emma Strebel.

The design specifically accentuates the experience of solar midday. Just as people enjoy watching the transition that takes place at sunrise and sunset, this design creates a space to observe midday: the moment when a particular place on Earth transitions from rotating toward the Sun to rotating away from the Sun. It disconnects viewers from time as an objective truth, and rather connects people to the inherent changes that are always occurring around us. The design specifically highlights one moment at one particular place on Earth. It develops a connection to the human biological need to experience changes in daylight. The design offers a meditative space to experience the dynamics in sunlight and learn about our profound relationship to earthly changes. While sunrise and sunset are inherently climactic changes in the experience of light throughout the day, midday often goes unnoticed. As the earth constantly rotates, no two consecutive moments of sunlight are the same. This design calls attention to the midday change— midday is the only moment at which the light hits the floor of the structure without hitting either of the east or west walls. Without this significant design element, midday would go unnoticed. Furthermore, the design highlights both solstices (when the beam of light reaches the southern and northern walls), which are also moments that would otherwise go unnoticed.

At any moment that someone encounters the design, it offers a visual comparison. For the first half of the day, the sunlight inches down the wall and across the structure. While the sunlight is constantly changing throughout the entire day, the moment of anticipation is for midday. The climactic moment builds up anticipation and offers a visual comparison for prior and following moments. Over the course of a year, this design frames the light in a way which highlights Earth's more gradual rotation towards and away from the Sun. It visually represents where the earth is on its journey around the Sun. Over the course of one year, the position of the sunlight starts at the south wall, moves toward the north wall, and then reverses to move back to the south wall. This element of the design accentuates seasonal changes in light. Experiencing these changes strengthens people's understanding of their ever-changing surroundings.

4 Discussion

This exploratory design is developed to meet the human need to experience natural change when indoors and build an understanding of environmental change at large. The aim is to create an experience that enables us to learn more deeply about the rhythms of daylight and thereby stimulate biological needs for connecting humans to nature. In doing so, viewers become familiar with the earth's timescale of change and learn about the interconnected systems we exist within [7]. Combining Lam [3] and Böhme's [18, 19] theories on biological needs and atmosphere, the design creates an atmosphere that tracks changes in sunlight in order to fulfill the biological need to connect to daylight. While creating a design, the goal is to illustrate how a tool can accentuate a moment of sunlight in order to help people notice a changing moment on Earth and learn about the relation between humans and the universe. The design detaches changes in sunlight from clock time as we know it and creates an awareness of the way sunlight moves through a space in order to accentuate the viewers' experience on Earth. By connecting to changes in sunlight, humans learn the importance of building a connection to their place on Earth.

This design concept can be implemented for site specific installations. The proposed design concept is a basic design that can be altered in many ways. In any scale or iteration around the world, the design offers an experience for learning more intimately about the relation between humans and the universe.

This paper proposes an exploratory design method to demonstrate how to use the geometry of the sunlight in order to create a spatial design that enhances the human connection to Earth and the natural rhythms of the day. In a broader context, this approach focuses on biological needs, developing an architectural space where experiencing the dynamics of the daylight connects us to our surroundings [3]. This method develops sustainable design grounded in one's natural surroundings in order to conceptualize the rhythms of the earth. These principles focus on experiential and exploratory learning [9] and can be integrated into architecture and lighting design beyond this particular project in order to investigate natural phenomena. Working with the geometry of sunlight at particular locations on Earth generates architecture that is better able to connect people to their surroundings. With a deeper connection to earthly changes, humans become more in tune with diurnal rhythms, and their connection to daylight and sunlight. In doing so, architecture and lighting design is able to enhance the human connection to light and "*set the internal world to the external world*" and visa versa [20]. With this fundamental connection, viewers build an understanding of the timescale of the earth's environmental changes and the interconnectivity of our place on Earth.

5 Conclusion

This paper sets up an exploratory design to investigate how to develop a lighting design rooted in the geometry of sunlight that connects humans to the rhythm of the day and year.

The design explores the movement of the earth and the position of the Sun in relation to one's location on Earth. By applying this natural scientific approach, the design creates a space centered around the experience of earthly changes in sunlight. It marks one moment of change at midday, the slower rotation of an entire day and the even more gradual fluctuations throughout the year. With these comparisons, viewers depart from their familiarity of clock time by connecting to visual changes and routines in sunlight. These connections help develop an understanding of the earth's constant fluctuations. Change has enabled our evolution, intertwined us with light and will continue to define our experience of life on Earth. As humans increase their impact on the earth's ecosystems, understanding its timescales has never been so important. This understanding is a step toward the human need to empathize with and take care of the environment at large. The design demonstrates how a design can be a tool that informs viewers about the complex and fundamental interaction between the earth, Sun and human. It develops a space to acquaint humans with the inner workings of sunlight, in turn connecting us to our place on Earth and in the Cosmos.

References

1. Rovelli, C.: The Order of Time. Riverhead Books, Translated by Erica Segre and Simon Carnell (2018)
2. Zajonc, A.: Catching the Light: The Entwined History of Light and Mind. Oxford Univ Press, New York (1995)
3. Lam, W.M.C.: Perception and Lighting as Formgivers for Architecture. Edited by Christopher Hugh. Ripman, Van Nostrand Reinhold (1992)
4. Vernon, M.D.: Perception through Experience. Methuen (1970)
5. Sagan, C.: Billions and Billions: Thoughts on Life and Death at the Brink of the Millennium. The Random House Publishing Group, New York (1997)
6. "Shifting Seasons." Conservation in a Changing Climate. climatechange.lta.org/climate-imp acts/shifting-seasons/. Accessed 20 July 2020
7. Meadows, D.H., Wright, D.: Thinking in Systems: A Primer. Chelsea Green Publishing, Vermont (2015)
8. Armstrong, S.J., Fukami, C.V.: The Sage Handbook of Management Learning, Education, and Development. SAGE Publications, London (2009)
9. "Learn Online With Us." Exploratorium. www.exploratorium.edu/
10. Strebel, E.: "Solar Midday." Aalborg University Copenhagen (2020). https://projekter.aau. dk/projekter/en/studentthesis/solar-midday(1467c423-27ca-4c14-99e9-d77ef42202d4).html
11. "Newgrange - World Heritage Site." Newgrange Stone Age Passage Tomb - Boyne Valley, Ireland. www.newgrange.com/. Accessed 20 July 2020
12. Turrell, J.: "Skyspaces." James Turrell. jamesturrell.com/work/type/skyspace/. Accessed 20 July 2020
13. McCaw, C.: "Sunburn." Chris McCaw. www.chrismccaw.com/sunburn. Accessed 20 July 2020
14. Pallasmaa, J.: The Eyes of the Skin. Wiley, Hoboken (2005)
15. Trotter, A.: In residence at Can Lis. Openhouse Mag. (2019). openhouse-magazine.com/jorn-utzon-can-lis/. Accessed 20 July 2020
16. Merleau-Ponty, M.: Phenomenology of Perception. Translated by Colin Smith, Routledge, London (1962)

17. Schön, D.A.: The Reflective Practitioner: How Professionals Think in Action. Basic Books, New York (1983)
18. Böhme, G.: Atmosphere as the fundamental concept of a new aesthetics. Thesis Eleven **36**(1), 113–126 (1993)
19. Ursprung, P., Böhme, G.: Atmosphere as the subject matter for architecture. Herzog & De Meuron: Natural History, Lars Muller, pp. 398–407 (2002)
20. Bragg, M.: "In our time." Audio Blog Post. Circadian Rhythms. BBC Radio, 17 December 2015. Web. 13 May 2020

Digital Games, Gamification and Robots

Intergenerational Playful Experiences Based on Digital Games for Interactive Spaces

Felipe Bacca[4], Eva Cerezo[4(✉)], Rosa Gil[2], Antonio Aguelo[1], Ana Cristina Blasco[3], Teresa Coma[3], and Maria Angeles Garrido[3]

[1] Departamento de Psicología y Sociología, Universidad de Zaragoza, Zaragoza, Spain
`aaguelo@unizar.es`
[2] Departamento de Informática e Ingeniería Industrial, Universitat de Lleida, Lleida, Spain
`rosamaria.gil@udl.cat`
[3] Departamento de Ciencias de la Educación, Universidad de Zaragoza, Zaragoza, Spain
`{anablas,tcoma,garridoa}@unizar.es`
[4] Departamento de Informática e Ingeniería de Sistemas, I3A Universidad de Zaragoza, Zaragoza, Spain
`ecerezo@unizar.es`

Abstract. In this article, we first review the work carried out in the field of intergenerational digital games experiences as well as in the identification of the design factors involved. They are valued according to their applicability to put a common point to generate Intergenerational playful experiences based on digital games for interactive spaces. Starting from that point, "The Fantastic Journey", a game created to be played in an interactive space where tangible interaction on tabletops, physical interaction with real objects as well as body and gesture interaction is supported, is valued as a possible intergenerational digital game experience. Two play sessions and a workshop carried out with grandparents and their grandchildren have allowed us to elaborate the findings in the literature about the potential and the factors around intergenerational play and have served to legitimize The Fantastic Journey as a true intergenerational digital game experience.

Keywords: Intergenerational · Hybrid digital games · Interactive spaces

1 Introduction

Older people represent a growing proportion of the world population. Between 2000 and 2050, the proportion of the world's population over 60 years will double from about 11% to 22%. In our occidental societies, the older adults often suffer from social and emotional isolation, and from ageism. The term 'ageism' has emerged to refer to both the negative attitudes towards older people, and the negative attitudes that older people hold towards young people. Studies [1–5] show that video games can be a cohesion tool that enhances socialization between young and old. In fact, digital games can be individually

---

F. Bacca—Deceased

© ICST Institute for Computer Sciences, Social Informatics and Telecommunications Engineering 2021
Published by Springer Nature Switzerland AG 2021. All Rights Reserved
E. I. Brooks et al. (Eds.): DLI 2020, LNICST 366, pp. 101–119, 2021.
https://doi.org/10.1007/978-3-030-78448-5_8

beneficial for both generations. For older adults, they can improve cognitive functioning (e.g., short-term capacity, memory, attention, hand-eye coordination) [6, 7]; overcome communication problems and social isolation [8]; and encourage physical exercising and prevent falls [9, 10]. For children, collaborative digital games can improve learning, skill building, and healthy development [11]; can encourage learning, exploration, experiment, and construction of knowledge; and can develop imagination and creativity [12]. Moreover, intergenerational interactions can be mutually beneficial for both collectives: breaking with some age stereotypes or ageist attitudes [13]; developing civic engagement and contributing to an age inclusive society [14]; linking the learning and leisure needs of both generations [14] encouraging communication, solidarity, and social connectedness between generations [10, 15].

In spite of these findings, the number of projects focusing in intergenerational games is scarce, also the works that focus in their design factors. One of the factors found in the literature is the prioritization of physical, mixed reality games and multimodal interaction as well as the convenience to use shared context and meeting places to enable social interactions. From that point of view, Interactive Spaces may play a role in supporting intergenerational playing experiences. Interactive Spaces (IS) are distributed user interfaces supporting several ways of interactions in digitally augmented rooms. They combine a panoply of related interaction paradigms such as Physical Computing, Context-Aware Computing, Mixed Reality, Wearables and Tangible User Interfaces, allowing multiple users to interact, at the same time or in a distributed way. The objective of the work presented here has been to explore the potential of Interactive Spaces to support intergenerational playing experiences. To do so, two game sessions with family groups, comprised of grandparents and their grandchildren, playing in an interactive space that supports tangible, gestural and body interaction, were carried out. We wanted to compare the findings of those sessions with the ones present in the literature. Moreover, we wanted to know if grandparents and grandchildren agree with the factors stated as fundamental in the literature to generate successful intergenerational playing experiences. This is why a third experience, an intergenerational workshop, was carried out. Paper structure follows.

Section 2 introduces the state of the art in the Intergenerational Digital Gaming literature, as well as the factors to take into account when designing intergenerational experiences. In Sect. 3 a game "The Fantastic journey" is presented and analyzed from the intergenerational perspective. In Sect. 4, the intergenerational playing sessions and the workshop carried out in the ETOPIA Art and Technology Center of Zaragoza's City Council are presented and analyzed. The conclusions section summarizes the experience and discusses future work.

2   State of the Art

We will first review the digital intergenerational games present in the literature and then the factors to take into account when designing intergenerational experiences.

2.1   Intergenerational Digital Gaming

Different types of digital intergenerational games can be found in the literature; some for family environments, sharing a location or through the internet, and others in which specific educational aspects are sought to be applied to populations of different ages, not necessarily among family members. After a general search about intergenerational games, the projects selected were those in which the target population is of extreme ages (children and the elderly, not necessarily family members) and in which any digital tool is used as an interactive medium to allow interaction whether required or not physical elements. The projects were classified according to the development technologies and predominant types of interaction (see Table 1).

Table 1. Development, technologies and predominant types of interaction

Project name	Location games with tangible interaction	Interactive physical experiences	Online Games	Games with experimental development
Curball [16]				X
Distributed Hide-and-Seek [17]			X	X
Age Invaders (Khoo et al. 2008) [18]		X	X	
Save Amaze Princess [15, 19]	X			
Atomium [9]	X			
Family Quest [20]				X
TranseCare [10]			X	
Xtreme Gardener [21]		X		
AR card game [22]	X			
Parent-Child Sexual Health Dialogue [23]	X			
Cooperative game to old powered chair users and their friends and family [24]		X		
Children's Museum [25]	X			
eBee [26]				X
Mr Robojump [27]	X			
Co-smonauts [27, 28]		X		
MeteorQuest [29]				X
SoundPlay [30]				X
Intergenerational shared action games [31]	X	X		

Looking at Table 1, we note a predominance of location games with tangible interaction; Save Amaze Princess [19], for example, essentially takes the game mechanics of Ludo and Snakes and Ladders and augments them with the use of a board projected onto a table and the use of physical tokens with animated feedback. This type of project is carried out on the hypothesis that, in general, an interactive intergenerational digital game is more successful if it takes as a reference traditional games already established, and if it also includes physical elements. It means that there is a lower learning curve of the game and a reduction in fear of technology for the elderly [25, 27].

On the other hand, Physical Interactive Experiences are preferred for exploring group interactions, typically of one to four participants. Age Invaders [18], for example, has a board in which people have to move to achieve their mission; at the same time they can interact online with other players. Xtreme Gardener [21], meanwhile, explores collaborative play to keep a garden protected from the elements and anything that can harm plants. The disposition of the game seeks that children and grandparents, through physical actions, control these elements. They are represented on a screen by their silhouettes that are tracked by a Microsoft Kinect device. Other experiences such as Cosmonauts [27] also resort to physical elements but not as controllers (tokens or symbols) but as playable pieces (parts of a rocket).

Online games are not usually predominant if we talk about intergenerational digital games, because the trend is that interactions invite people to share. However, it should be noted that Age Invaders [18], for example, adds this possibility to enrich the participatory game dynamics: grandparents and grandchildren on the one hand, and parents through the internet. Distributed Hide-and-Seek [17] may be a very interesting bet on online gameplay, since it does not have grandparents and grandchildren sitting remotely in front of a screen. Considering the impossibility, many times, of being together they propose to play a physical game such as "hide and seek" with the help of sensors so that children can hide and be found, in an entertaining way, by adults in a defined space.

Finally, Experimental development games consider the use of various technologies to verify their effectiveness. eBee [26] uses an entire previous dynamic of co-creation with weaving grandmothers to create small hexagonal pieces woven by hand with electronic components. These pieces make a board dynamic highly attractive for children due to the various colors and textures. The game pieces were made by grandmothers, manually, with crochet stitching, as a symbolic form of cohesion and identification with the activity. MeteorQuest [29], a ubiquitous game with mobiles, proposes that by means of geolocation, the family travels through certain areas in a city "hunting" the fragments of a meteorite that has fallen to the ground. This experience, enriched with creative dynamics, encourages teamwork to solve the clues required in the search and like no other, takes digital intergenerational games off the wall.

As it can be seen from the previous works, physical interaction, or at least mixed virtual/physical experiences, and co-located playing seem to be important factors for successful intergenerational games. From that point of view, Interactive Spaces (IS) [32] may be a natural place to deploy intergenerational games. Initially, ISs have been applied to explore new possibilities of collaborative work and meeting rooms but their use to support ludic experiences [24] is also rapidly growing. Nevertheless, in order to explore their potential to support successful intergenerational games the identification of the factors to be considered in their design is needed.

2.2 Design Factors of Intergenerational Experiences

Regarding the factors to be taken into account to design intergenerational digital games, the work of De la Hera et al. [4] stands out. In their work, they make an exhaustive review of the state of the art in that moment in order to obtain, not only information about the benefits of intergenerational digital game-playing practices, but of the design factors to be taken into account. They group benefits around three important questions: strengthening of family ties, reciprocal learning and greater mutual understanding and reduction of social anxiousness. They find that the way it has proven to be most effective at narrowing the gap between generations and motivating mutual learning is through narratives used as the basis for game mechanics design. Related to the design factors, they classify them into two types of factors that are important to take into consideration: player-centric and game-centric factors. In Table 2, the factors and the related findings in the works analyzed are summarized.

The work of Kolthoff et al. [5] is also relevant. From the work of De la Hera et al. [4] and the works of Chiong [33] and Zhang and Kaufman [2], they propose 13 design factors shown in Table 3 including their applicability. Kolhoff et al. [5] contrast these theoretical factors found in previous studies with interviews among elderly and youth. Interviews confirmed the importance of five factors (weighing of different motivations young and old; need for a learning component; options for a short game; ease of use and communication and nature of social interaction) and added that the game has to be funny and save about terms of privacy.

Having detected the potential of interactive Spaces to support intergenerational gaming experiences and the most important design factors to consider, we decided to make use of a previously created Interactive Space [35] to follow that research line.

Table 2. Factors and findings in intergenerational digital games (elaborated from De la Hera et al. [4].

	Factors to consider	Findings
Player-centric	The nature of interactions between older (51–81 years old) and younger (4–22 years old)	• Users tend to carry out asymmetric interactions, where grandparents act as grandchildren "supporters". At the same time, grandchildren want to be considered as "skilled students" by their grandparents • Interactions use to build from histories inspired by the grandparents

(continued)

Table 2. (*continued*)

	Factors to consider	Findings
	The motivations to play digital games	• Both grandparents and grandchildren seek relaxing and having fun. Grandparents also seek social interaction and a way to escape from their reality • Children like long games, whereas old people prefer shorter games • Grandparents prefer avoiding games related to reflex movements (running, fighting…).They have more difficulties in those kinds of games and they do not enjoy them so much. They avoid violent games • Grandparents adapt to the game's content much better than young gamers do. In this way, maybe, it is interesting to design games according to young people's preferences
	The difference in habilities	• Due to their physical and cognitive difficulties, old people may have difficulties in understanding and using the games' devices depending on the technology used • Enactive interactions, which are not based on specific digital competences or mental models, are a good solution to deal with the differences in abilities • Children may also have trouble when technology is not adjusted to their age and abilities
Game-centric	Goal-related forms of interaction	• Older gamers tend to be less competitive and assume a more passive or supportive role • Better results are obtained if there is a collaborative competition: competitive games with a collaborative background promote the interaction between old and young people
	Space related forms of interaction	• Interaction mechanisms work better if they are carried out in presence of other people, participants or spectators (co-locative) • In the case of VR games, extra communication functionalities, such as sound and touch are welcome: they facilitate older participants to interact and motivate children as they can teach them how to use them

Table 3. Intergenerational games design factors and their applicability (elaborated from Kolthoff et al. [5])

Design factors	Applicability
Weighing in different motivations from both age groups	There must be motivations for both types of user to make the game attractive
Learning embedded in the game	All approaches must include some aspect related to learning
Short game sessions	Both age groups prefer games with short gaming sessions
Easy to use interface	The interface should be simple and easy to use for both groups
Collaboration games with common goals have best fit for both	Collaborative games should focus on joint goals and avoid competition between the two participants (elderly-young), but rather competition with other teams or the system
Peer-to-peer mentoring by teaching each other	The design of the system should encourage reciprocal learning
Enable social interaction, shared context and meeting places	It is important that social interaction arises with not only the participants of the game, incentivizing competition and empowering participants; to achieve this, co-locative experiences and spaces where it can be socialized outside the experience should be used
Video chat and computer mediated communication helps	In the case of experiences in different locations, communication is decisive
Asymmetrical and asynchronous play	Asymmetric learning (in which both users do not supplement the same role but are fed back) and turn-based play is more conducive to this type of game
Nature of interaction in important	Interaction must be conceptually in some common term between young and old
Enable passive watching play	Allow the person watching the game (usually the elderly) to also get satisfactory feedback
Prioritize physical, mixed-reality games and multi-modal interaction	Performing actions in the space allows multiple people to participate in the interaction at the same time
Create socially desired reward systems	This is relevant in virtual games or gamified group interactions where there are additional incentives and activities that do not necessarily involve the game

3   Playing in an Interactive Space: The Fantastic Journey

After studying the factors and recommendations to support successful intergenerational playing experiences we realized that a pervasive game, The fantastic journey, previously developed, could be a good starting point to support that kind of experiences.

3.1   Game Description

The fantastic journey is a game, initially designed to work attention, planning and social skills with ADHD children, developed by the AffectiveLab Group at the University of Zaragoza with the support of educators and therapists [34]. The game has been designed to be played in the JUGUEMOS Interactive Space [35]. It is an indoor space of around 70 m^2 that includes a real-time localization system, two Kinects (to support gesture interaction), microphones, and projectors. It also includes a set of four NikVision tangible tabletop devices which have been proved to be useful for kids to improve their cognitive, manipulative and social skills [36].

The fantastic journey is an adventure game, in which the protagonist has to progress over the story interacting with different characters and objects. It is a multiplayer game, aimed up to 16 players organized in four groups. The game is intended for use by inexperienced players that do not need to be familiar with the rules of the game neither with the controls. During the play, players have to move around the interactive space in order to find objects or to achieve the challenges proposed by the game. They also have to interact with the tangible tabletop devices and use their own body or voice. The groups have to collaborate to help the main character complete the journey, and is articulated around several missions commented next.

Magic words. Here, players have to pay attention to the lyrics of a song and then, order the words that make up the chorus. It is made with physical words put on the tabletop devices.

The sun and the moon. In this mission, players have to make up the shapes of the sun and the moon (projected on a wall) by placing themselves (localization) inside the silhouettes.

The search for the suitcase. Here, players have to find a suitcase hidden in the IS. The suitcase is closed with a padlock. The key can be obtained by playing Starloop [37], a game that were developed to improve computational thinking in kids (see Fig. 1).

Fig. 1. Intergenerational group playing Starloop.

Keyword. This mission allows working attention in both selective and global levels. Children will listen to a story in which a word is constantly repeated. Then, they will have to find the word in a word search that is projected on the tabletop devices (see Fig. 2left).

Fig. 2. Word search (left). Planet of Indians (right).

Planet of Indians. Here, players have to follow sound patterns, so successive processing and selective attention are worked. Each tabletop device represents a color and a sound (see Fig. 2right). Players have to reproduce a sound sequence by hitting the tabletop with drumsticks in the correct order.

Freeing the stars. Here, the goal is to free three stars that have been trapped in a spider web. Selective attention and simultaneous processing are the abilities to develop. The player has to select the elements required by means of gestural interaction (see Fig. 3).

Fig. 3. Freeing the stars.

Meteorite attack. This mission is about destroying a set of meteorites. It helps to work on selective attention and planning of time-space paths. The meteorites get destroyed by shooting them with spaceships on the tabletops (see Fig. 4left).

Butterflies. In this game, players must stay quiet so that the butterflies that are projected on the walls are placed on the flowers and can be counted. The idea is to work on the inhibition of impulsive behaviors and on self-control.

Fig. 4. Meteorite attack (left). Encounter with the Comet of laughs (right).

Encounter with the Comet of laughs. The last phase of the game consists of a projection of the last scene, in which the protagonist meets the Comet of laughs (see Fig. 4right), and of the playing of the song of the game, which will be sung and danced by the players to celebrate the success of the mission.

3.2   The Fantastic Journey as an Intergenerational Experience.

The game fulfills several of the factors presented in previous section (see Table 3) to be important to support intergenerational playing experiences. In particular:

- Learning embedded in the game: each activity/mission has been defined to work one or more competences (such as language, learn to learn, social skills, digital competence and competence of initiative).
- Short game sessions: the game has been structured around short missions that are solved and allow it to continue in the game; nevertheless, a common and engaging narrative drives the experience which will better fit expectations from the younger gamers (see also Table 2).
- Easy to use interfaces: natural interaction based on manipulating objects using the tangible tabletops (magic words, keyword, planet of indians, meteorite attack activities), seeking of objects within the space (search of the suitcase activity), or using the own body (the sun and the moon activity) or hands (freeing the stars activity) avoiding the interaction through specialized or complex devices.
- Collaboration games with common goals have best fit for both: although teams are formed to play, they have to collaborate: all the teams have to achieve the goal so that they can all continue to the next activity; this promotes interaction not only within a team but among teams.
- Enable social interaction, shared context and meeting places: the Interactive Space acts as a meeting place that allows co-located play.
- Prioritize physical, mixed-reality games and multi-modal interaction: the game supports tangible, gestural and bodily interaction.

To increase both groups' motivation and engagement with the narrative (in our game not originated from elderly stories) we decided to include a greeting from Pipo's grandma video to welcome the families when they entered the Interactive Space (see Fig. 5).

Moreover, in order to assure easy to use for all, and taking advantage of the fact that during play, the game is controlled by a mediator, it was decided to allow the mediator to choose for each mission the level of difficulty (each mission has been designed with different levels of difficulty) that better fit.

Fig. 5. Getting to know Pipo's grandma (left) and Pipo main character (right)

In the next section, the intergenerational experiences carried out based on the fantastic journey game are presented.

4   Intergenerational Experiences

The JUGUEMOS interactive space is located in the ETOPIA Art and Technology Center of Zaragoza's City Council where families are engaged all year long in different artistic/ technological activities. We decided to organize game sessions where children with their grandparents could play The Fantastic Journey together. The objective of the sessions was to assess the potential of the game to support intergenerational play, getting direct feedback from users, and to observe the dynamics of the intergenerational groups in the interactive space to compare them with the findings in the literature. Two sessions were carried out. After them, it was decided to organize an intergenerational workshop to deepen with the families in the intergenerational games design factors. All the precedent mission figures were taken during those experiences.

4.1   Intergenerational Game Sessions

Two intergenerational game sessions were organised, one in December 2018 and the other June 2019. They were announced through the municipal web and family groups formed by one or two grandparents and one or two grandchildren aging from 7 to 12 could sign up. In the first group there were 18 people so two families were put together in one of the tabletops. In the second group, there was just a family in each tabletop. One researcher took observation notes and two others helped players with the different missions. The families played the game for around one hour. Afterwards, players were divided into two age groups: grandparents filled a questionnaire and children just talked about the experience. The family groups were able to get through all the missions without special difficulties. Compared to usual children-only groups, children were observed to be more quiet and careful when playing. Following main observations are commented.

With regard to participation:

In general, they all helped and facilitated that all of them could see what was happening and could participate in the missions. They all celebrated their achievements. Especially grandparents celebrated them, singing Pipo's song after the missions.

In terms of leadership:

Children made proposals and grandparents observed or helped. In general, the grandparents were much more prudent than children, acting slowly, leaving the children to make decisions and act. When children failed to perform activities, then grandparents began to act and to make decisions.

Regarding mediation:

Grandparents provided the children with the materials needed for the missions. They gave instructions and advice when children got stacked. Some grandparents organized turns among their grandchildren and encouraged them to help other teams after having accomplished their own missions. In some activities, interaction was quite intensive:

- In the Starloop mission, some grandparents participated quite actively, giving advice about the best strategies. They also expressed curiosity asking the children to explain how they had succeeded in completing the activities ("Why have you put this tab here?")
- In the suitcase mission, the suitcase was a quite old-fashion one and the grandparents had to explain to the children how to open it.
- In the word search, grandparents got involved much more actively giving precise instructions.

During the small talk with the children after playing, many of them said that even though they spend quite a lot of time with their grandparents they do not play with them ("It has been the first time my grandfather has been playing with me"). They all thought that their grandparents would not be able to finish the game without their help. Nevertheless, they admitted that in some missions, such as the word search, their grandparents had been better than them.

In the questionnaire, grandparents were asked about their feelings playing with their grandchildren and the difficulties encountered by both of them. The answers were very positive showing a general very positive experience. Anyway, two issues arose. Surprisingly, they felt they had not helped their grandchildren as they had seen them very good at playing, which was not always the fact. This may indicate the necessity of pointing out the value of the supporting role of the elderly during the game play. Besides, regarding the feelings they had felt during the play, although the most common terms were happy/very happy, the words slow and stupid arose. This points out the necessity of carefully tuning all the activities to the abilities of all the participants, which represent a big challenge as they may be very different even among individuals of the same age, as it is also the case among children.

Regarding the observations if we compare them with De la Hera et al. [10] findings (see Table 1), we find some agreements but also disagreements:

- "Users generally tend to carry out asymmetric interactions, where grandparents act as supporters for grandchildren". This was observed during the whole game. In fact, the game mediator had to encourage the elderly to take a more active role in the missions.
- "Grandparents adapt to the game's content much better than young gamers. In this way, maybe, it is interesting to design games according to young people's preferences." It is true that elderly adapt well to all types of games or missions (for example to tangible tabletop activities) but also the children to "more adult" missions (the word search). In addition, we realize that using a game based on a familiar activity for the elder allows them to show themselves as "masters", what they love, and, as the children saw them as experts, it helps to break down usual age stereotypes. On the contrary, those activities in which the elderly see that children are more used (using a Kinect in the Freeing the stars mission) make them assume a more passive role (the mediator had to encourage the elderly and just one of them took part).

The experience was very positively considered in both age groups. They all agreed they had had a good time and thanked the opportunity of playing together. We realize that, although we split the age groups to comment on the experience, they were very interested in commenting on the experience with the other age group. In fact, we realize that talking about the game experience itself could be a new good conversation topic for them and could also foster intergenerational interaction. We wanted also to discuss with them the most important factors too take into account when designing intergenerational games. Therefore, in the next experience, we decided to shorten the play experience and to add other intergenerational activities, as explained in the next section.

4.2 Intergenerational Workshop

In this second type of experience, four family groups participated, made up of one or two grandparents (over 60 years old) and one or two grandchildren (under 12 years old) without cognitive or motor difficulties. The experience consisted of three parts. First, they played a simplified version of the game with only three activities: Meteors attack, Starloop and Freeing the star. They were used because they combined fun (Meteors attack) and learning (Starloop) and tangible (Meteors attack and Starloop) and gesture (Freeing the stars) types of interaction. The design factors brainstorming (part 2) and the games modifications proposals (part 3) are explained next.

Design Factors Brainstorming
The aim of this part was to talk with grandparents and their grandchildren about the most important factors to design intergenerational games to see if they agree with the ones mentioned in the literature.

The activity consisted of interpreting, taking into account the clues, a message in Japanese, given by a grandmother, who was also a game designer (see Fig. 6 left). This activity also sought to intensify intergenerational participation and to took advantage of the previous gaming experience to decipher four aspects to take into account when designing games to play by grandparents with their grandchildren.

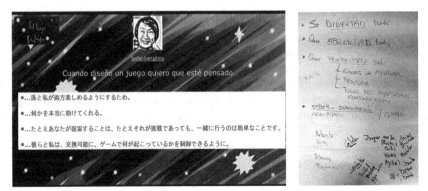

Fig. 6. Message form the grandma designer in Japanese (left). Brainstorm written in a flip chart (right)

The four aspects the Japanese grandmother argues to consider, are:

A. The game might allow both my grandchildren and me to have fun.
B. It has to be useful for something, and above all, so that my grandchildren and I are left wanting to spend more time together, playing.
C. What game proposes, even if it is a challenge, must also be something with the possibility of doing it together, and in turn, it must be as easy for my grandchildren as it is for me.
D. Both they and me, indistinctly, can have control of what happens in the game, without the game requiring personal resources that exceed one or the other.

In fact, the four proposed aspects are related to the model proposed by Cheng [38] that integrates the technology acceptance model (TAM) and the theory of planned behavior (TPB).

The technology acceptance model (TAM) [39] considers that:

– Perceived ease-of-use refers to the extent to which an individual believes that using a particular system is free of effort.
– Perceived usefulness refers to the extent to which an individual believes that using a particular system would improve work performance.

The theory of planned behavior (TPB) [40] considers that:

– Attitudes toward the behavior refers to an individual's favorable or unfavorable response to a particular behavior. It should be noted that the original model [41], in addition to the aspects considered by Davis (utility and ease of use), also included the Self- Esteem and the Subjective norms (individual's reaction to social preferences on performing a particular behavior).
– These authors also consider the importance of the Perceived behavioral control, key element in relation to intergenerational games.

To build the four aspects that the supposed Japanese designer takes into account, the three variables of the integrated model (TAM+TPB) of Cheng [45] that were of special interest for this workshop were considered: The perceived usefulness (A & B aspect proposed by the designer), perceived ease-of-use (C aspect) and perceived behavioral control (D aspect). Furthermore, the four aspects are related to most of the design factors proposed by Kolhtoff et al. [5], as can be seen in Table 4, except those that include online aspects not applicable to this experience.

Table 4. Kolthoff et al. [5] design factors with the four factors worked in the workshop

Design factors	Aspects related with...
Weighing in different motivations from both age groups	A,B, C and D
Learning embedded in the game	B
Peer-to-peer mentoring by teaching each other	B
Enable social interaction, shared context and meeting places	B
Create socially desired reward systems	B
Collaboration games with common goals have best fit for both	C
Prioritize physical, mixed-reality games and multi-modal interaction	C
Easy to use interface	C, D
Nature of interaction in important	C, D

A brainstorm was carried out and the ideas were written in a flip chart (see Fig. 6 right). It should be noted that the group arrived to principles very similar to those of the Japanese designer, highlighting the fun, learning (utility), participation of all (something easy to do together) and control over the game.

Games Modification Proposals
In the third activity, the Starloop and Freeing Stars games were evaluated, based on the four principles derived from the previous activity (see Fig. 7). The participants were divided into two intergenerational groups and each one analyzed one of the games, and then, shared what to keep, what to eliminate, what to change, and what to add.

Fig. 7. Presenting the proposed game modifications (Starloop)

The improvement proposals, above all, were aimed at suggesting small changes in the experience, without proposing significant changes in any way. The key proposals focused in particular on "utility" (B) and "ease of use" (C). This coincides with the two points that have special relevance for the TAM model and for the items of the selected model of Kolthoff et al. (in Table 4 most of the elements have a connection to B and C).

Suggestions to improve the experience affected the following aspects:

- Changing colors (more squeaky colors in the stars to make them more fun;
- Simplification of processes (introduce the possibility that the tours to the stars of the star game Starloop, could also be programmed diagonally);
- Inclusion of learning elements (related to science, the universe, the stars, incorporating the incidence of gravity in the Meteorites attack mission);
- Increase of the level of complexity of the task so that the games offer more possibilities (including some more galactic stars to the game Freeing the stars).

Regarding the game in general, a child literally expressed: "Make it longer and more complex, with more tests". This goes in the same direction stated by De la Hera et al. [10]: younger gamers like longer and more challenging experiences. This question has to be carefully considered as making the experience longer or more difficult could affect the elderly experience. The use of physical controls ("I have always liked video games with physical controls" expressed one of the elderly) and bodily interaction ("movement in the air") were also welcomed.

5   Conclusions and Future Work

Intergenerational ludic experiences may have important benefits for both collectives and can contribute to increase mutual understanding, but are still scarce in the literature. Physical interaction and co-located play appear in the literature as two important factors for successful intergenerational interactions. Both aspects can successfully be supported in Interactive Spaces where groups of different ages may interact and have fun and learn together.

A game designed to be played in a public interactive space supporting physical, tangible, gesture and body interaction has been used to carry out two play sessions and a workshop with grandparents and their grandchildren. The experiences have been analyzed and have confirmed the suitability of the designed games to support intergenerational play and have also helped us to fine tune the design factors and recommendations found in the literature.

In spite of the positive results, the experiences have brought to light some questions:

- the necessity of strengthening the learning potential of the ludic experiences;
- the utility of the game experience as a new conversation topic that may facilitate the dialog between generations.
- the potential of such experiences to overcome prejudices between generations showing different roles and abilities;
- the challenge of fine tuning the game to the cognitive and physical abilities of all participants;

– the lack of tools to assess those intergenerational experiences and their impact.

Future work will focus on those issues, in particular, on how to potentiate dialog between generations and on the design of specific assessment methods to evaluate the short, mid and long term impact of the experiences.

Acknowledgments. Work partly funded by the Spanish Science, Innovation and University Ministry (MCIU), the National Research Agency (AEI) and the EU (FEDER) through the RTI2018-096986-B-C31 contract and by the Aragonese Government (Group T60_20R). This project has received funding from the European Union´s Horizon 2020 research and innovation programme under the Marie Sklowdowska-Curie grant agreement N° 801586 and from the University of Zaragoza.

References

1. Newzoo: Male and Female Gamers: How Their Similarities and Differences Shape the Games Market. https://newzoo.com/insights/articles/male-and-female-gamers-how-their-similarities-and-differences-shape-the-games-market/. Accessed 11 Sept 2020
2. Zhang, F., Kaufman, D.: A review of intergenerational play for facilitating interactions and learning. Gerontechnology **14**(3), 127–138 (2016). https://doi.org/10.4017/gt.2016.14.3.013.00
3. Costa, L., Veloso, A.: Being (grand) players: review of digital games and their potential to enhance intergenerational interactions. J. Intergenerational Relat. **14**(1), 43–59 (2016). https://doi.org/10.1080/15350770.2016.1138273
4. De la Hera, T., Loos, E., Simons, M., Blom, J.: Benefits and factors influencing the design of intergenerational digital games: a systematic literature review. Societies **7**(3), 18 (2017)
5. Kolthoff, T., Spil, T.A., Nguyen, H.: The adoption of a serious game to foster interaction between the elderly and the youth. In: 2019 IEEE 7th International Conference on Serious Games and Applications for Health (SeGAH), Kyoto, Japan (2019)
6. Whitlock, L.A., McLaughlin, A.C., Allaire, J.C.: Individual differences in response to cognitive training: using a multi-modal, attentionally demanding game-based intervention for older adults. Comput. Hum. Behav. **28**(4), 1091–1096 (2012)
7. Cota, T.T., Ishitani, L., Vieira, N.: Mobile game design for the elderly: a study with focus on the motivation to play. Comput. Hum. Behav. **51**, 96–105 (2015). https://doi.org/10.1016/j.chb.2015.04.026
8. Allaire, J.C., McLaughlin, A.C., Trujillo, A., Whitlock, L.A., LaPorte, L., Gandy, M.: Successful aging through digital games: socioemotional differences between older adult gamers and non-gamers. Comput. Hum. Behav. **29**(4), 1302–1306 (2013)
9. Abeele, V.V., De Schutter, B.: Designing intergenerational play via enactive interaction, competition and acceleration. Pers. Ubiquit. Comput. **14**(5), 425–433 (2010). https://doi.org/10.1007/s00779-009-0262-3
10. Derboven, J., Van Gils, M., De Grooff, D.: Designing for collaboration: a study in intergenerational social game design. Univ. Access Inf. Soc. **11**(1), 57–65 (2012). https://doi.org/10.1007/s10209-011-0233-0
11. Thai, A.M., Lowenstein, D., Ching, D., Rejeski, D.: Game changer: investing in digital play to advance children's learning and health. The Joan Ganz Cooney Center at Sesame Workshop, New York (2009)

12. Bredekamp, S., Copple, C.: Developmentally appropriate practice in early childhood education. National Association for the Education of Young Children, Washington DC (1997)
13. Harwood, J.: Understanding Communication and Aging: Developing Knowledge and Awareness. Sage, New Delhi (2007)
14. Uhlenberg, P.: Introduction: why study age integration? Gerontologist **40**(3), 261–266 (2000)
15. Mahmud, A., Mubin, O., Shahid, S., Martens, J.B.: Designing and evaluating the tabletop game experience for senior citizens. In: Proceedings of the 5th Nordic Conference on Human-Computer Interaction: Building Bridges, pp. 403–406. ACM (2008)
16. Kern, D., Stringer, M., Fitzpatrick, G., Schmidt, A.: Curball–a prototype tangible game for inter-generational play. In: 15th IEEE International Workshops on Enabling Technologies: Infrastructure for Collaborative Enterprises (WETICE 2006), Manchester, pp. 412–418 (2006). https://doi.org/10.1109/WETICE.2006.27
17. Vetere, F., Nolan, M., Raman, R.: Distributed hide-and-seek. In: OZCHI, pp. 325–328 (2006). https://doi.org/10.1145/1228175.1228235
18. Khoo, E.T., Cheok, A.D., Nguyen, T.H.D., et al.: Age invaders: social and physical inter-generational mixed reality family entertainment. Virtual Reality **12**, 3–16 (2008). https://doi.org/10.1007/s10055-008-0083-0
19. Mahmud, A., Mubin, O., Shahid, S., Martens, J.B.: Designing social games for children and older adults: two related case studies. Entertain. Comput. **1**(3–4), 147–156 (2010)
20. Siyahhan, S., Barab, S., Downton, M.: Using activity theory to understand intergenerational play: the case of Family Quest. Int. J. Comput.-Support. Collab. Learn. **5**, 415–432 (2010). https://doi.org/10.1007/s11412-010-9097-1
21. Rice, M., Yau, L.J., Ong, J., Wan, M., Ng, J.: Intergenerational gameplay: evaluating social interaction between younger and older players. In: CHI 2012 Extended Abstracts on Human Factors in Computing Systems (CHI EA 2012), pp. 2333–2338. Association for Computing Machinery, New York (2012). https://doi.org/10.1145/2212776.2223798
22. Lin, C.-L., Fei, S.-H., Chang, S.-W.: An analysis of social interaction between older and children: augmented reality integration in table game design. In: Holzinger, A., Ziefle, M., Hitz, M., Debevc, M. (eds.) SouthCHI 2013. LNCS, vol. 7946, pp. 835–838. Springer, Heidelberg (2013). https://doi.org/10.1007/978-3-642-39062-3_64
23. D'Cruz, J., et al.: Promoting parent-child sexual health dialogue with an intergenerational game: parent and youth perspectives. Games Health J. **4**(2), 113–122 (2015). https://doi.org/10.1089/g4h.2014.0080
24. Seaborn, K., Pennefather, P., Fels, D.I.: A cooperative game for older powered chair users and their friends and family. In: Proceedings of the 7th IEEE Games Entertainment Media Conference (IEEE GEM 2015), pp. 52–55 (2015). https://doi.org/10.1109/GEM.2015.7377242
25. Dietmeier, J., Miller, B.J., DeVane, B., Missall, K., Nanda, S.: Shredding with mom and dad: intergenerational physics gaming in a children's museum. In: FDG 2017: Proceedings of the 12th International Conference on the Foundations of Digital Games, August 2017, pp. 1–4 (2017). Article No: 58. https://doi.org/10.1145/3102071.3106365
26. Carlsson, I., Choi, J., Pearce, C., Smith, G.: Designing eBee: a reflection on quilt-based game design. In: Proceedings of the 12th International Conference on the Foundations of Digital Games (FDG 2017), pp. 1–10. Association for Computing Machinery, New York (2017). Article No: 24. https://doi.org/10.1145/3102071.3102102
27. Lankes, M., Hagler, J., Gattringer, F., Stiglbauer, B.: InterPlayces: results of an intergenerational games study. In: Alcañiz, M., Göbel, S., Ma, M., Fradinho Oliveira, M., Baalsrud Hauge, J., Marsh, T. (eds.) JCSG 2017. LNCS, vol. 10622, pp. 85–97. Springer, Cham (2017). https://doi.org/10.1007/978-3-319-70111-0_8

28. Lankes, M., Hagler, J., Gattringer, F., Stiglbauer, B., Ruehrlinger, M., Holzmann, C.: Cosmonauts in retrospect: the game design process of an intergenerational co-located collaborative game. In: Proceedings of the 2018 Annual Symposium on Computer-Human Interaction in Play Companion Extended Abstracts, ACMDL, pp. 221–234 (2018)
29. Rosenqvist, R., Boldsen, J., Papachristos, E., Merritt, T.: MeteorQuest - bringing families together through proxemics play in a mobile social game. In: Proceedings of the 2018 Annual Symposium on Computer-Human Interaction in Play (CHI PLAY 2018), pp. 439–450. Association for Computing Machinery, New York (2018). https://doi.org/10.1145/3242671.324 2685
30. Mushiba, M.: SoundPlay: an interactive sound installation for playful intergenerational encounters in public areas. In: AfriCHI 2018, pp. 1–3 (2018). https://doi.org/10.1145/328 3458.3283506
31. Seaborn, K., Lee, N., Narazani, M., Hiyama, A.: Intergenerational shared action games for promoting empathy between Japanese youth and elders. In: 2019 8th International Conference on Affective Computing and Intelligent Interaction (ACII), pp. 1–7 (2019)
32. Jetter, H.-C., Reiterer, H., Geyer, F.: Blended Interaction: understanding natural human-computer interaction in post-WIMP interactive spaces. Pers. Ubiquit. Comput. 18(5), 1139–1158 (2013). https://doi.org/10.1007/s00779-013-0725-4
33. Chiong, C.: Can video games promote intergenerational play & literacy learning. In: Report from a Research & Design Workshop. The Joan Ganz Cooney Center at Sesame Workshop, New York, vol. 1, pp. 8–12 (2009)
34. Gallardo, J., López, C., Aguelo, A., Cebrián, B., Coma, T., Cerezo, E.: Development of a pervasive game for ADHD children. In: Brooks, A.L., Brooks, E., Sylla, C. (eds.) ArtsIT/DLI -2018. LNICSSITE, vol. 265, pp. 526–531. Springer, Cham (2019). https://doi.org/10.1007/978-3-030-06134-0_56
35. Bonillo, C., Marco, J., Cerezo, E.: Developing pervasive games in interactive spaces: the JUGUEMOS toolkit. Multimed. Tools Appl. 78(22), 32261–32305 (2019). https://doi.org/10.1007/s11042-019-07983-6
36. Marco, J., Baldassarri, S., Cerezo, E.: NIKVision: developing a tangible application for and with children. J. UCS 19(15), 2266–2291 (2013)
37. Marco, J., Bonillo, C., Cerezo, E.: A tangible interactive space odyssey to support children learning of computer programming. In: Proceedings of the 2017 ACM International Conference on Interactive Surfaces and Spaces, pp. 300–305. ACM (2017)
38. Cheng, E.W.L.: Choosing between the theory of planned behavior (TPB) and the technology acceptance model (TAM). Educ. Tech. Res. Dev. 67(1), 21–37 (2018). https://doi.org/10.1007/s11423-018-9598-6
39. Davis, F.D.: Perceived usefulness, perceived ease of use and user acceptance of information technology. MIS Q. 13(3), 319–340 (1989)
40. Ajzen, I., Fishbein, M.: The influence of attitudes on behavior. In: Albarracin, D., Johnson, B.T., Zanna, M.P. (eds.) The handbook of attitudes, pp. 173–221. Erlbaum, Mahwah (2005)
41. Ajzen, I.: The theory of planned behavior. Organ. Behav. Hum. Decis. Process. 50(2), 179–211 (1991)

Discourses of Digital Game Based Learning as a Teaching Method

Design Features and Pedagogical Opportunities Associated with Teachers' Evaluation of Educational Game Apps

Jeanette Sjöberg[1] and Eva Brooks[2(✉)]

[1] Halmstad Universty, Kristian IVs väg 3, 301 18 Halmstad, Sweden
jeanette.sjoberg@hh.se
[2] Aalborg University, Kroghstræde 3, 9220 Aalborg, Denmark
eb@hum.aau.dk

Abstract. In recent years, digital games have increasingly become an important part of children's lives. As a consequence, digital game-based learning (DGBL) activities have also been merged into the school context and tried out by teachers in various ways. The pedagogical and didactical values of integrating DGBL in education are however not yet concluded. In this paper we examine how groups of teachers construct ideas about digital game-based learning as a teaching method and base for developing teaching activities. The study is drawn from a couple of creative workshops with Swedish and Danish school- and preschool teachers, in which their pedagogical design processes while evaluating and trying out different game apps have been studied. The research questions we ask in this paper are: 1). In what ways do teachers concretise their comprehension of digital game-based learning in their discussions of educational games for school children? And; 2). How are different discourses about the learning process and/or didactical potential in relation to digital games constructed in teachers' discussions while assessing game apps? Using a discourse analytical approach, the results of the study show that the teachers' were stuck by their preconceptions about games as offering different learning qualities compared to their traditional teaching practice. Teachers acknowledged that DGBL is a complex issue as also designers' preconceptions are tied to traditional qualities of game design.

Keywords: Digital game-based learning (DGBL) · Discourse analysis · Educational game apps · Game-based design · School teachers · Teaching method

1 Introduction

Digital games are an important part of children's lives. Arguments stating that games for learning can be used to provide authentic, effective, and joyful educational experiences are well documented [1–4]. By integration of learning content into games, researchers have put forward that such digital game-based learning, or DGBL, could have potentials for being motivating and engaging as well as promoting students' achievements [5, 6].

E. I. Brooks et al. (Eds.): DLI 2020, LNICST 366, pp. 120–139, 2021.
https://doi.org/10.1007/978-3-030-78448-5_9

However, the actual impact of DGBL has as well been questioned. Kickmeier-Rust and Albert [7], point to how a poor design of educational games can influence the learning process and outcome. The authors emphasize the key issue of carefully considering the learning design aspect when designing educational digital games, for example by providing learning guidance to balance the relationship between a game's gaming and learning aspects. Nevertheless, game design factors often are an overlooked matter [8, 9], for example, game mechanisms, game goals, and game narratives, as well as the way a learner interacts with and controls a game (keyboard, joystick, motion-sensing) can impact learning-related aspects [10]. The issue of how the design of a game can support learning is however less studied [11].

Beside considering game design features as crucial aspects of DGBL, it is important to point to another topic in this field of research that has gained limited attention, namely a focus on teachers and their implementation as well as facilitation of DGBL. This still remains a challenging issue [12, 13]. The present paper presents outcomes from a bigger study on DGBL in a Nordic context, including the countries of Denmark, Finland, Iceland, Norway, and Sweden. In this project, our findings show that the reason to why teachers do not apply digital games in their teaching primarily relates to that they experience a lack of pedagogical and technical capacity to decide in what ways and when digital games or gamification tools would apply to their teaching goals [14]. Relating this to the above-mentioned issue of game design shortcomes, Squire [15] points to game-based learning as a two-sided problem, where, on the one hand, game designers can develop inspiring games, but are less competent to design for games to support educational activities. On the other hand, teachers are knowledgeable about qualities of teaching material, but do not know so much about what kind of design features that make a digital game effective to use. The present paper has taken these aspects into consideration by including 12 teachers in a workshop to investigate how different discourses emerge when they are offered tools to assess digital games' pedagogical as well as design values and, further, to design a lecture including game-based learning. Through this, our intention was to study the process of teachers' discussion while assessing and designing for DGBL. It was our hope that this approach also could serve as a resource to facilitate their further understanding, awareness, and implementation of DGBL in their teaching activities.

The following sections start with a research overview of DGBL as a learning method, followed by a specification of the research questions. Next follows a description of the theoretical framework, including analytical tools for identifying design features of games supporting DGBL. This is followed by the methodological framework detailing the method design. Finally, we present the outcomes of the study followed by a concluding discussion.

2 Digital Game-Based Learning as a Learning Method

Arguments concerning beneficials and effects of using digital games for learning have increased tremendously among researchers and educational practitioners in recent years [16, 17]. Moreover, DGBL is increasingly highlighted as a contemporary alternative and effective way to develop learning. Or, as Papadakis expresses it: "Digital games are gaining wide recognition as an effective way to create socially interactive and constructivist learning environments" [18]. The various game components that are part of DGBL, such as competition, commitment, instant reward and feedback, are elements that individually as well as together are considered beneficial for learning [16–20]. Contemporary games are developed to satisfy basic requirements of learning environments and can thus provide an important tool in supporting teaching and learning processes. Based on Van Eck [20], Nousiainen et al. [21] have pinpointed four different game-based learning approaches, namely: (1) using educational games, (2) using entertainment games, (3) learning by making games, and (4) using game elements in non-game contexts. In the same study, the authors identified possible competence areas needed by teachers to work with DGBL [21]. Their results showed that teachers' DGBL capacities should be more integral to their professional knowledge and skill repertoires. Hence the authors bring to the fore the importance of developing more knowledge about DGBL focusing on teachers' learning [21]. In our study we further this research by taking hold of facilitating teachers' awareness and professional development of DGBL. In doing so, we primarily focused on Nousiainen et al.'s first approach, where the participating teachers in our study used educational games while exploring digital game apps.

A major point regarding the effectiveness of including games in learning processes has to do with the principle of situated cognition and the fact that the learning takes place within a meaningful context [20]. Since the subject matter is directly related to the learning environment, the gained knowledge is both applied and practiced by the learner. Van Eck states that "games are effective in learning not because of what they are, but because of what they embody and what learners are doing as they play a game" [20]. However, in order to successfully integrate educational methods and game design, there is a need for an in-depth understanding of the various possibilities that digital games might provide [17]. In order to fully capture what games have to offer, Plass et al. [22] claims that a combination of cognitive, motivational, affective, and sociocultural perspectives is necessary for both game design and game research. Even though many positive claims have been made about DGBL, there are, as previously mentioned, some sceptical opinions towards using games as educational tools [7, 23]. Critics question the viability of DGBL and argue that research has been slow to provide empirical evidence on its effectiveness. Van Eck writes:

"Scepticism about games in learning has prompted many DGBL proponents to pursue empirical studies of how games can influence learning and skills. But because of the difficulty of measuring complex variables or constructs and the need to narrowly define variables and tightly control conditions, such research most often leads to studies that make correspondingly narrow claims about tightly controlled aspects of games" [20].

Another side to this has to do with the idea that research within the area still is scarce [24, 25]. One such area is related to teacher knowledge [26–29]. Hébert and Jenson [29] argue in their study that there are little to none research that examines either teacher pedagogies in relation to digital game-based learning, or professional development for teachers on DGBL either focus on pedagogy or study the impact of professional development on teacher practice. Conclusively, DGBL as a learning method is not yet totally defined.

In this paper we examine how groups of teachers construct ideas about DGBL as a teaching method and base for developing teaching activities. To do this we followed their pedagogical design process while they evaluated and tried out different digital game apps. The research questions posed in this paper are as follows:

1. In what ways do teachers concretise their comprehension of digital game-based learning in their discussions of educational games for school children? and;
2. How are different discourses about the learning process and didactical potential in relation to digital games constructed in teachers' discussions while assessing digital game apps?

3 Design Features Supporting Digital Game-Based Learning

Inspired by Shi and Shih's literature review [8] focusing on higher level game design concepts that are not restricted by genre, we have identified what we consider as essential when designing for DGBL in school settings. These concepts form the analytical lens of the present study. As a core concept of game design features, which all design factors should be based upon, Shi and Shih [8] assign game goals. These kinds of goals should provide learners with certain gaming experiences to inspire them to, for example, explore game content and also for them to experience satisfaction of achieving goals of the game. Hartmann and Klimmt [30] point to that gaming achievements, in general, means that the learner gains some kind of power, gathers game objects, or competes with others. Game goals from a learning point of view stipulate how such experiences may relate to curricula to support specific learning goals. In other words, game goals resonate with teaching objectives and the experiences these objectives are supposed to provide for the learners. The overall game goals can be divided into three categories, game mechanisms, game fantasy, and game value [8] and will be further elaborated in the below text.

Game mechanism refers to how the game enables a learner to smoothly navigate in the virtual game world. This means that the learner, through interacting with the game, can experience how it is triggered to generate relevant feedback [8]. From a learning point of view, the game's interface becomes a crucial design feature, for example by displaying hints on the screen and providing game characters that can assist and guide the learner throughout the game. In other words, the design feature of interaction determines the learner's gaming process and provides feedback and, in this way, allows for the learner's autonomy. This kind of autonomy enables the learner to, for example, create, select, and change, increasing his or her sense of engagement [31]. As such, game mechanisms and its subcategories of interaction and autonomy are influential to learning processes [32].

Game fantasy involves a game's environment and background. From a game-based learning perspective this means that elements of the game must be integrated into an imaginary world, where the learner becomes immersed in the game. This means that the learner's experience is closely related to the game fantasy feature. This is also where the game's narrative becomes a key as it describes what happens in the game world [8]. For games that target learning, narrative is important to provide the learner with informative knowledge. However, Shi and Shih [8] and Hoffman [33] highlight the importance of having the teaching aspects well connected into this rather than being added and thereby disconnected. Thus, teaching content should match the narrative to establish a game that in a meaningful way supports learning [34]. Sensation constitutes another aspect of a game's fantasy and, based on the narrative, it refers to audio and aesthetics, such as graphical elements, which is supposed to increase the learner's motivation [35].

Game value attracts learners to start playing the game and represent special features that only exist in the game and as such they are the reason why a game is experienced as joyful [8]. Expressed differently, game value is a core factor for learners to generate motivation and engagement [36]. To obtain game value, the learner achieves rewards by, for example, managing tasks and challenges, and reaching goals. Challenges must be meaningful for the learner to generate game value and should be considered in regard to the learning objectives that are in focus as well as to the learners' skills [8, 32]. Sociality as a game feature is vital to nurture collaborative learning. A game can be designed for sociality through, for example, its interface to support communication or competition between learners. This is done through a game's mechanism [8]. In other words, sociality needs to be designed to elicit learners' collaborative activities.

In the context of the present study, we have involved primary school teachers, preschool teachers/leader, and a teacher educator to evaluate digital games targeting learning in classroom settings. We have asked them to consider different games' learning objectives, interface design, aesthetics, game mechanisms, and game values. These are vital factors of DGBL for teachers to choose a specific game and for learners to enjoy learning while playing a digital game. This and other methodical issues will be further elaborated in the following text.

4 Method

The present study is based on qualitative research including two creativity workshop cases (Case 1 and Case 2) designed to provide a framework for primary- and preschool teachers to assess potentials of digital game-based learning. Accordingly, a number of selected apps in the areas of math, language, and science were introduced to the participating teachers aligned with an assessment framework to value the apps' learning designs in terms of content and form. They were divided into groups to choose a game and to jointly evaluate this game.

Case 1 included nine female primary school teachers from schools in south-west Halland, Sweden. The nine teachers (three from preschool and six from primary school) were divided into three groups (two in group 1, four in group 2, and three in group 3). The group of four teachers included teachers from the same school and teacher team. The remaining groups included participants from different schools. *Case 2* included three male participants, a preschool teacher, a leader of preschools, and an assistant professor in a teacher education programme focusing on mathematics. In addition, a female toy- and game designer participated in Case 2. The Case 2 participants worked together in one group.

Each group had a designated workstation where a fixed camera facing the table centre was set up and recorded the activities at each workstation. In total, we gathered 400 min of video data. Additional 80 min of video data from Case 2 were lost and, therefore, the four participants in Case 2 were interviewed a while after the workshop to capture their further insights on the topic of DGBL in classroom settings. In addition, the empirical data consist of the participants' final presentations as well as field notes by the two authors.

4.1 Apparatus

The teachers received some background material before starting the game app workshop. First, they got a general introduction to game-based learning, for example that it is not a new phenomenon, but has been around for decades. Chess, for example, was used in the middle Ages to teach strategic thinking. Further that the origin of preschools, mid 19th century, was based on Friedrich Fröbel's ideas about learning through games and play. In addition to this, the introduction included some general information about game mechanics and their implications in a learning context, for example that a game-based approach is based on rules, clear goals and includes choices that end up with different consequences. A game designed for learning is supposed to offer opportunities for teachers and students to collaborate around specific game contents and in this way add depth and perspective to the student's gaming experience. However, even though students, in general, spend lots of time on digital game play, this does not automatically mean that they appropriate the learning that teachers have assigned to a DGBL session. Finally, the teachers were introduced to the purpose of different kinds of games, for example winning games, achieving goals games, collaborating games, explorative games, and problem-solving or strategizing games. After this introduction, the teachers were divided into groups and started the workshop activities.

4.2 Procedure

The workshop was divided into four parts, where in particular parts two and three beside a research goal also targeted to serve as a method that the participants should be able to use also after the workshop. This is due to the fact that our previous study [14] showed that teachers ask for knowledge about and framework for assessing teaching and learning values of digital games. Thus, our method applied in part two and three of the workshop included questions about the game's design as well as its teaching and learning potentials, i.e. considering a combination of both game and learning designs. Table 1 illustrates the design of the workshop.

Table 1. Workshop design

Time	Activities
14:00–14:15	Introduction of the workshop and selected apps. Establishing the workshop framework and climate
14:15–14:30	Workshop part 1: Exploring the different game apps
14:30–15:20	Workshop part 2: Playing and assessing the chosen game app focusing on game design and teaching and learning potentials
15:20–16:10	Workshop part 3: Development of a teaching activity by means of the chosen game app
16:10–17:00	Workshop part 4: The groups present their resulting teaching activity for each other arguing for their design choices. Closing and evaluation of the workshop

The introduction of the workshop consisted of clarifying definitions of DGBL as well as the goal of the workshop. Moreover, the chosen game apps were presented to the participants. Considering our previous study [14] where we identified that digital games primarily were used in the fields of mathematics, language, and science, these became the areas of apps for the present workshop. Tables 2 and 3 describe the specific games in Case 1 (Sweden) and Case 2 (Denmark).

Table 2. Game apps used in Case 1 (Sweden).

Swedish language	Mathematics	Science
Spelling game (Stavningslek)	Math bakery 1, 2, and 3 (Mattebageriet)	Chemist
School writing (Skolstil)	Critter Corral	Twitter (Kvitter)
Letter puzzle (Bokstavspussel)	Scratch Jr.	Butterflies (Fjärilar)
Yum letters (Yumbokstäver)		

Table 3. Game apps used in Case 2 (Denmark).

Danish language	Mathematics	Science
Leo & Mona Reading fun (Leo & Mona Læsesjov)	GOZOA - Play and learn mathematics (GOZOA - Leg & lær matematik)	The hero of nature (Naturens helte)
The letter school (Bogstavskolen)	Pixeline - The labyrinth of the number master (Pixeline - Talmesterens labyrint)	
	Scratch Jr.	

While Case 1 included a mixture of digital games and digital tools, Case 2 included only digital games.

In the workshop part 1, the participants had time to test the different apps and choose one of them to assess as well as to design a teaching activity. This was followed by a longer session (workshop part 2) where they should more in detail play and assess the game design to get ideas about how the game could be used for a specific teaching activity. This part of the workshop was assisted by a list of questions to guide the evaluation:

- What is the goal and value of the game - is it clear and pedagogically convincing? Why or why not? What kind of learning goals can the game cover?
- The game interface - is it easy and efficient to navigate?
- What are the rules, control and other mechanisms of the game? How can the player learn and understand those rules and mechanisms?
- Is the game balanced by offering different game levels? If so, in what way?
- What kind of mechanisms or values would encourage the child to play it again?

- In what way has the game an aesthetic value?
- What kind of game - Is it based on exploration, problem solving, contesting, or a mixture?
- In what way is the game engaging, motivating?
- As a pedagogical expert, would you use this app in your teaching activities? Why or why not?

Part 3 of the workshop had designated time for the participants to develop a teaching activity which should be based on the chosen app. Here, they did not get any guidelines but were told to apply their pedagogical knowledge and competence, in particular related to the learning goals that would apply to the chosen game (Fig. 1). This was followed by workshop part 4, where the participants presented their digital game-based teaching activity design for each other and argued for the included choices, game design features as well as pedagogical benefits and/or restrictions (Fig. 2).

Fig. 1. Participants from the Danish case developing DGBL designs.

The participants were informed about the study in writing and agreed to having the workshop sessions video recorded by signing informed consent forms. In line with ethical guidelines, all names of the participants as well as of their workplaces are anonymized, and accordingly no identifying information is provided.

Fig. 2. Participants from the Swedish case presenting their DGBL designs.

4.3 Data Analysis Approach

The methodological approach used in our analysis of the video recordings originates from discourse theory [37–39], and partly from discursive psychology [40, 41]. Within discourse analysis, language use is formed in social contexts and viewed as a tool by which people construct the social world [37–39]. These processes are performed in a non-mechanical and heterogeneous way, and according to Fairclough [38], numerous discourses coexist and contrast each other as well as compete with one another in various social domains. When it comes to discourse analysis, language and language relations are referred to each other in understanding the reality of social actions, where the individuals are the actors who produce the discourses. The focus can be both on micro- and macro perspectives. In the present study, however, the focus is on the micro level since it revolves around teachers' specific reasoning and their way of seeing and understanding the reality in which they live, in contrast to a macro perspective that rather would illuminate a larger area of the society. From this perspective, constructions of discourses should be examined out of the assumption that discourses jointly created leads to certain positions and actions are made possible, while other positions and actions are made impossible within a specific practice, an assumption that stems from a social constructionist perspective [42]. The basis of social constructionism is to study the general relationship between man and society based on language as a significant and central tool, which means that reality is socially constructed and that people through language construct their own world [42]. Wetherell [43] emphasizes that discourse is something that inspires and is supposed to have a good foundation to stand on, but it is also provocative and difficult to interpret. To assist the analysis process we followed five analytical steps (see Table 4). To get an overview of the empirical material, the video data was transcribed verbatim (step 1). This was followed by step 2 where we, through colour coding, identified discourses in the material. To identify recurring patterns of constructions, we next carried out a joint review of the data (step 3). Out of the identified patterns we ordered them into themes and analysed excerpts in line with discourse structure (step 4). Finally, as step 5, we chose representative examples from the excerpts and decided which of them to include.

Table 4. Analytical steps in the discourse analysis

Steps	Activities undertaken	Foci guided by analysis
Step 1	Transcription of video recordings, total material	Overall view of the material
Step 2	Colour coding of specific content	Identifying discourses in the material
Step 3	Joint review to discern patterns	Identifying recurring patterns of constructions
Step 4	Organization of themes out of patterns	Analysing excerpts in accordance to discourse structure
Step 5	Selection of representative examples of excerpts	Deciding which excerpts to include

To further help us in our analytical process we have used a couple of discourse analytical concepts: *interpretation repertoire* and *constructions*. The purpose of interpretation repertoire is to understand how humans and the world around them are constructed in connection with social actions and interactions [44]. The analysis thus focuses on how interpretive repertoires are built up and understood in a specific context through language as action - by analysing language and how interpretive repertoires are built up and maintained, the presupposed knowledge is thus challenged [42]. When it comes to the concept of constructions, there is a basic principle within discourse psychology where man is considered to be able to construct different versions of the same event and the same phenomenon [44]. This can be regarded as a consequence of language being constructed in different ways and it is thus not considered problematic that stories and arguments can vary in a text. In this study, constructions refer to the teachers' varied ways of talking about the same thing - the phenomenon of digital gaming apps - and how they present these variations in their discussions.

Having analysed the data, it was possible to - after the visualization of patterns and linguistic expressions in the teachers' discussions - identify three emerging themes in the material: *game design as persona, game design as form* and *game design as pedagogical function*. These themes, presented below, are connected to different perspectives of the concept of game design and should be seen in relation to the aim and research questions in the present paper.

5 Results

Theme 1: Game Design as Persona

This theme is related to the discourse constructed in the teachers' discussions about the game's design and the way in which they talked about it. Fairclough, [38] who theoretically positions himself between the structural and the socio-cultural perspectives, describes this discourse as a discourse practice. This should, on the one side, be understood as a way in which game designers produce texts, and, on the other hand, representing a socio-cultural practice, i.e. how players pick up or use the game design. In this case, the discourse is constructed by the teachers as they reflect on the text (i.e. the game app). This is to say that the game design itself (e.g. what kind of choices it offers, the mechanics, whether it is a single or multiplayer game, and so on), and when the teachers pick up this game design, they move between text and social construction or action. When the teachers talked about the game apps, the border between the text and sociocultural aspects blurred in terms of games/apps becoming personalized. This is to say that the text was not only framed as a text but over-layered with subject-like signs. A common feature that emerged in the recorded material is that the participants often refer to the game app as "it" or "they" and in several cases in relation to personal characteristics, as is exemplified below in Excerpt 1.

Excerpt 1:

Case 1, group 2. In this example, the four teachers in this group have individually been trying the game app 'Math bakery' (a math game app) for a while, and are now discussing their experiences of that as well as the advantages and disadvantages with the game app in relation to learning.

> Teacher 1: *If you move the cookies, you get results that are shown on the number line in a clear way...*
> Teacher 2: *So, yes, it is not totally dumb...*
> Teacher 3: *Should I show mine too? I think it is clear, to...* [she points to the screen] *...here we train multiplication, here I choose...different kinds of cookies, so here I actively choose which table I want to train on. Then rungoes on as you also have with stars and so on. And here it's great, here they show the different ways...*

In this excerpt, the teachers refer to the game as being 'clear' and not (being) 'totally dumb', which are human qualities that they attribute to the game itself. In addition, teacher 3 also refers to the game app as 'they' ('they show'), which also points to how the teachers construct the game's persona. In doing so, the teachers construct a discourse of the game in which the game is presented as a subject rather than an object. In applying this kind of personalized attributes to the game app, it is worth noting the teachers reproduced a discourse producing a kind of hybrid 'thing', which was inflected by embodied attributes. This blended conceptualization of the game app is, however, problematic as it produces an indeterminate way of talking about game apps as this is not distinct enough to draw attention to the issue of using games in learning activities. It is worth noting that themes 1 and 2 spanned between the teachers' views on a game app as a persona and the game apps' design, i.e. what they communicate, offer, and what is possible to do with them. This, in turn, leads to the inherent pedagogical opportunities (theme 3). Thus, next follows theme 2 focusing on the teachers' perspectives on what the game apps' design offered.

Theme 2: Game Design as Form

A game's structure enables learners to navigate and interact in a game, i.e. tied to a game's goal, this structure forms the game's rules offering the player different choices to navigate throughout a game. In this way, goals, rules and choices are tied to more complex procedures that all in all form the content of gameplay. However, unlike other media, the form of digital games still does not have an established structure, for example, they can vary in their mechanics, narrative, scope, topic, or number of players. This can create difficulties for teachers to determine what it is that constitutes a game, in particular when it comes to educational games. Considering this, the theme 'game design as form', addresses the issue whether a game actually is a game or not, and it is constructed by the teachers' discussions about what a game is, or what it is that constitutes a game. As they are talking (sometimes almost negotiating) about what really makes out a game, they are constructing a discourse of the game in which they are positioning the game app as either a (real) game or as something else (for example a puzzle or a pedagogical tool). This is exemplified below in Excerpts 2–4.

Excerpt 2:
Case 1, group 1. There are two teachers in this group and they have chosen the game app 'Letter puzzle' (a language app focusing on spelling progression) and are discussing the qualities of it in relation to the characteristics of what makes out a game. Their discussion is guided by the questions handed out in the beginning of the session.

> Teacher 1: *The purpose* [with the game] i*s that the letter sounds can be sounded together, into words.*
> Teacher 2: *Yes.*
> Teacher 1:*... and connect words and pictures also maybe. That this is the next step to writing. 'What kind of game is it?'* [Teacher 1 reads from the sheet of paper with questions].
> Teacher 2: *Well, you get rewards when you're right, the balloons .... but there are no clear rules.*
> Teacher 1*: Is it a game at all when there are no rules?*
> Teacher 2: *You mean if this is really a game?*
> Teacher 1: *So, I would not really call it a game, it depends on how you define it. It's more of a puzzle.*

In this excerpt the two teachers are reasoning whether or not the game app really is a game or not. They are highlighting different criteria for what constitutes a game, such as the fact that you get rewards when you are right and that there are no clear rules. The last criterion, however, seems to weigh heaviest and they agree that the app does not really live up to what constitutes a game since there are no clear rules.

Excerpt 3:
Case 1, group 3. Here, the three teachers in this group are discussing the app 'Scratch Junior', which is more of a programming app than a game app - it is not a game in itself,

but it admits people to make and play games with it. They are discussing what the game app is actually about, what you can do with it, and whether or not it actually is a game.

> Teacher 1: *It is probably more problem solving… But there are no given problems. It is not the case that you go into the app and have to solve different problems and advance to different levels. That is not the case.*
> Teacher 2: *And you should not collect points or… It is more like an educational tool. Perhaps more that than a game.*
> Teacher 3: *If we choose problem-solving, it is very clear. The problem is being formulated.*
> Teacher 2: *Yes, you decide the problem yourself.*
> Teacher 3: *Then it can become very clear to those who will work with it. That we do it so that we can solve this or that.*
> Teacher 1: *Mmm. You add a purpose.*
> Teacher 2: *But if you think about it, is it a game?*

Here, as in the previous excerpt, the teachers are referring to fundamental criteria for what makes a game (e.g. when teacher 1 talks about advancing to different levels and teacher 2 talks about collecting points). Thus, the teachers are constructing a discourse of the game which contains a common understanding of what (really) constitutes a game, namely the game mechanics. Consequently, this discourse also implies a clear marking of what does not constitute a game. According to the teachers, if a game does not have a clear goal or gameplay and rules it is not a game.

Excerpt 4:
Case 2, group 1. Like in excerpts 2–3, the teachers talk about game mechanics, but in more general terms compare to the previous excerpts. They firmly state that the digital games for young children are based on simple mechanics and, even though the technique is available, they do not offer the needed aesthetics or explorative narratives to be regarded as a real game. Similar to the two previous excerpts (Excerpt 2–3), they discuss this in relation to game criteria. As detailed in the method section, case 2 teachers participated in a follow-up interview, excerpt 4 is an extract from this interview.

> Teacher 1: *It is a challenge to find good games that not only focus on learning, but also have explorative opportunities. Most of what we find includes that the child shall manage a level in a game and if you do not manage it, then, it is just a pity. You have to find something else to do. This creates a bit of an A and a B team of game players. If you cannot manage a level, you are out and not part of the playing team. Beside this, you cannot be curious about something in these kinds of educational games. A game consists of rules, that's how it is, you cannot be curious about something, I mean, on something that you jump into while playing.*
> Teacher 2: *Something that we discuss a lot, in relation to how, that you on the one hand have the necessary technique [to develop games that are more explorative] and, on the other hand this about right or wrong answers or choices when you play this kind of game. And if you transfer this to pedagogical thinking, then we come to that while playing this kind of game the child will do something right or*

wrong. And the more you make the wrong choice or answer wrong on a question in relation to what is expected from the game design, the less explorative you become. You'll stop exploring. What we lately have talked a lot about in relation to level-based games is what is called sandbox-games. This kind of game offers exploration for you to take your own initiatives towards what you yourself think would be exciting to do or explore. There are no right or wrong answers. Not anything that needs to be solved in a certain way. If you cannot solve it you leave it to another time and move on. Unfortunately, there are not so many games in this genre. They are coming though. But where they are coming is in relation to adult players, not children.

Teacher 1: *Yes, that's right. It is like this. In relation to technical issues, there are many high quality, complex game alternatives for adult game players, but if we look at it in relation to children, these games are simple, very simple. Regardless what game you choose. There are no details like in adult games. So, children miss out on this extra dimension, the aesthetics. Adult players can be involved in aesthetically designed games, but not children.*

The three examples in excerpts 2–4 combined show how the discourse of what a game really is emerges in the teachers' discussions and how they, through their speech, construct a truth about games in which games are defined. While excerpts 2 and 3 acknowledge the mechanics of a game design to decide whether a game is a game or if it is something else, excerpt 4 highlights the aesthetics and narrative of a game as crucial for meaningful games. The teachers emphasize that if a game does not offer exploration and ignites curiosity, the game becomes simplistic in its game design and children lose their interest and curiosity, i.e. a game is more than its mechanics.

Theme 3: Game Design as Pedagogical Function

This theme focuses on the teachers' interpretation repertoire and construction of how they can go about using DGBL, i.e. how the game design functions pedagogically. The ways teachers talk about the usage of games' in a classroom context put forward their collaborative, practical, and subject appropriate functions as foundational. The teachers emphasize that the collaborative function is not built into the game design, but rather needs to be designed by the teachers as an additional function outside the games. The practicalities related to the usage of game apps refer to that there is not a tablet for each child, which means that there might be several children using the same tablet. This hinders children's progression in the game. Referring to Burr [42], this should be understood in relation to when the teachers on the one hand consider that digital games can support pedagogical functions (e.g. learning math) and sometimes not (e.g. fostering progression in learning math). In addition, teachers are strongly bound to curriculum, both when it comes to content and progression. In this way, the teachers position themselves and create interpretive repertoires based on their own pedagogical beliefs or on the surrounding institutional context.

Excerpt 5:

Case 1, group 2. In this excerpt the teachers are talking about how to introduce their game app to their students in a teaching activity. The three of them have tried out the game apps Math bakery 1–3, and are discussing how these apps can be integrated in the learning context.

> Teacher 1: *For our third graders, we would say that here you have the opportunity to rehearse differently, because here* [in math bakery 1] *you do not have to go through line-up and such, but if it had been new, you would have had to talk about how to set up ... and have a lesson first, or if you have never worked with multiplication before. Then you would have had to go through it. But multiplication is not put in the hands of someone who has not done it before.*
>
> Teacher 3:... *if you have a lesson and say 2x6 or 6x2, it does not matter because it is the same. I think it's good here* [in Math bakery 3], *it explains a lot, you can clearly see that it does not matter.*
>
> Teacher 1: *You need to connect it to a smartboard and show them* [the school children], *or that you as an adult explain. So they know what they can get out of it. Otherwise it will just be like, now you can play a little, that they focus on the game.*
>
> Teacher 2: *Here you want them to test, so they can see how to line up.*
>
> Teacher 1: *Yes, but then you have to show them and explain.*

In this excerpt, the teachers highlight in what way the game app can be introduced in a meaningful way depending on the previous experiences of the children. This put forward the importance of the teachers role. The scenario that unfolds here is that if the teacher does not take the lead, the DGBL will be 'just like playing around' rather than instructional learning. Another matter that the teachers' discussion draws attention to is the material process of solving mathematical problems. Here, they underline that the game is clear and properly explains that 2×6 and 6×2 are the same. These examples show how the pedagogy of the game design is constructed as traces of social practice.

Excerpt 6:
Case 2, group 1, where the excerpt is taken from the follow-up interview with the teachers. In this extract teacher 1 is concerned that children in fact do not learn anything from the game apps that currently are available, i.e. their pedagogical function does not exist. Teacher 2 agrees and acknowledges that learning always happens, in particular if the focus is on the process rather than on a specific end product.

> Teacher 1: /.../ *what it means when you can discover and find your own way, what does it mean when you actually can learn something from it based on your own curiosity? The point relates to the existing culture around these games. The most important is to challenge the concept around learning. There is too much learning that is stupefying. And that is if we only look for the correct result, we risk to miss what else is around us. Take for example mathematics. Many researchers say that children don't handle mathematics, they just have skills to count. They have not learnt to understand math as a concept. /.../ And it is the same with game apps for children. We really want to teach them something, but we focus too much on the end result and forget to give the children opportunities to understand the surrounding world.*

Teacher 2: *Yes, one doesn't necessarily need to focus on learning - it comes as a bonus, no matter what. There are more opportunities for learning if one doesn't focus on learning. It's a paradox.*
Teacher 1: *There is another kind of game that is not so apparent within game apps for children, namely social games. Where many children can play at the same time. Where they can explore together. /.../ That's fascinating* [to do]. *Educational games for children don't have a child perspective. /.../ The kind of sandbox games, for example, there are games* [for adults] *that have inbuilt physical laws like when building a bridge, if it is not correctly done, it will fall apart. And I can try out another solution /.../ These kinds of games would inspire children to learn by collaborating or discovering.*

This example illustrates that the pedagogical function in game apps for children does not exist. The core is how the teachers' interpretative repertoire is inflected by the socio-cultural context within which they are situated. This context is based on regulations and perspectives that highly acknowledge pedagogical aspects such as collaboration, learning through exploring, acknowledging children's interest and curiosity. Thus, this excerpt draws attention to the significance of the sociocultural framing, both in questioning the meaningfulness of existing educational game apps as well as questioning the culture of game designs as such.

Excerpt 7:
Case 1, group 1. In this excerpt, the teachers are comparing two language apps focusing on spelling progression ('Spelling game' and 'Letter puzzle') in order to choose one of them to work with. They are talking about the difficulties of the games in relation to the children's level of knowledge.

Teacher 1: *Letter puzzle, the one with the sheep, we have that, we have used it quite a lot.*
Teacher 2: *Yes, because this one* [refers to the app she plays, 'Spelling game'] *still feels a bit complicated. I think like this, that when they* [the school children] *already have a hard time spelling and so the letters are hidden too...*
Teacher 1: *Yeah.*
Teacher 2: *... so that you do not even see, you must first search for the letters...*
Teacher 1: *Yes...well, it's both, it can be that they think it's a bit fun too, that it will be... for the other is very simple, if you say you have the letters there, you only have to put the pieces together.*

This excerpt exemplifies how the teachers construct an interpretive repertoire about their common understanding of DGBL and how it can be used pedagogically in the classroom. In this example, their interpretive repertoire is based on the adaption of the game app in relation to the school children's knowledge level. Teacher 2 is concerned that one of the game apps might be too hard for the school children, while teacher 1 points out that it must not be too easy for the school children, and states that what is difficult can also be fun. In all, excerpts 5–7 constitute different examples of how the teachers are constructing an interpretive repertoire which reflects a socially constructed ideology about learning and how it is (best) supported for enhancement in a classroom context.

To sum up, these three themes illustrate different perspectives of the concept of game design. The first theme, *game design as persona*, shows how game attributes became personalized by the teachers by conceptualizing them as 'hybrid things' with embodied qualities. The second theme, *game design as form*, acknowledges teachers' views on what it is that makes a game to a game. Here, on the one hand, game mechanics, such as clear goals and rules, were qualities that decided if a game was a game or something else (e.g. a puzzle). On the other hand, it was highlighted that a game is more than its mechanics and that aesthetics and narrative of a game are crucial aspects if a game should be a game. Finally, the third theme concluded that the *pedagogical function of game design* is constructed as traces of social practice. This was articulated partly through teachers beliefs in their instructional tradition and partly through a critical approach to current educational game apps as too simplistic in their design resulting in the fact that a meaningful pedagogical function does not exist.

6 Conclusive Discussion

By following four groups of teachers' pedagogical design processes while they were evaluating and trying out different game apps, we wanted to examine how ideas about DGBL as a teaching method and base for developing teaching activities were constructed by them. We wanted to investigate the teachers' views on the didactic potential of DGBL. In addition, our intention was to offer the participants a structured form to contemplate games with didactic eyes. Hence, while exploring the game apps together with colleagues they were establishing an organized way of evaluating and implementing DGBL in their teaching activities.

Related to the first research question ("In what ways do teachers concretise their comprehension of DGBL in their discussions of educational games for school children?"), the analysis showed that the teachers' were stuck by their preconceptions about games as offering different learning qualities compared to their traditional teaching practice. They tried to put the game apps into a pedagogical framework and their discussions concerned how they could be used for activities that they used to implement. Thus, they identified limitations of the game apps as they did not supply traditional activities. However, the teachers try hard to understand the principles of games, but run into problems when they find that they are not completely compatible with traditional methods. This confirms related work [15] highlighting that teachers are competent when it comes to qualities of traditional teaching material, but not so much about how digital games can foster learning. Squire [15] also points to another problematic side of DGBL, namely that designers are knowledgeable in developing inspiring games, but have limited knowledge about designs for learning. Related to the second research question ("How are different discourses about the learning process and/or didactical potential in relation to digital games constructed in teachers' discussions while assessing game apps?"), one of the groups pointed to that designers' limited competence of games for learning also influenced the games' design resulting in games with little or no relevance for teaching activities. In this regard, it is arguable that DGBL concerns more than an effective game mechanics, which is confirmed by related work [8], which points to the importance of game narrative and game value in terms of, among others, sociality.

6.1 Implications for the Field

We already know that, for instance, Nordic teachers for various reasons do not use DGBL to any great extent in their teaching [14]. The most evident reason is that they lack the knowledge about both the gaming context itself and about what it might enable in terms of learning potential, hence more knowledge about DGBL is needed for teachers to increase the use of games in school teaching. Accordingly, we argue for the necessity to view DGBL as an alternative method to be considered based on its specific attributes rather than approaching it in terms of traditional ways of teaching. Otherwise there is a risk that digital gaming aspects of learning disappear into a fuzzy pedagogical framework and the gaming aspects thereby become as fuzzy. This is to emphasize that the two-sided problem expressed by Squire [15] risks to become even more complex considering that it might be that both teachers' and designers' competences are tied to traditional qualities of pedagogy respectively game design instead of putting on new spectacles and consider DGBL for what it is rather than something traditionally framed.

We found that our workshop design encouraged a discussion and insight into games and their learning potential among the participating teachers- which it looked like they did not initially have - and we argue that our workshop design actually facilitates the use of DGBL as well as reduces the gap between games and pedagogy. Furthermore, the game apps used in our workshop design were chosen on the basis of what types of games people usually use [14] which we wanted the teachers to choose from, based on their own interest. This was related to the idea that they, in the workshop, should design a teaching activity which included the chosen game app and which they were going to carry out in their respective classroom. The next step to further this research is to perform follow up-interviews with the participating teachers in order to investigate how their teaching activities actually unfolded in the classroom.

Acknowledgements. The authors would like to thank all the teachers who participated in the research. The data were collected within the project 'Nordic DGBL - Digital Computer Games for Learning in the Nordic Countries', supported by Nordplus Horizontal, NPHS-2016/10071.

References

1. Hamari, J., Shernoff, D.J., Rowe, E., Coller, B., Asbell-Clarke, J., Edwards, T.: Challenging games help students learn: an empirical study on engagement, flow and immersion in game-based learning. Comput. Hum. Behav. **54**, 170–179 (2010)
2. Dede, C.: Theoretical perspectives influencing the use of information technology in teaching and learning. In: Voogt, J., Knezek, G. (eds.) International handbook of information technology in primary and secondary education, vol. 20, pp. 43–62. Springer Science + Business Media, New York (2011). https://doi.org/10.1007/978-0-387-73315-9_3
3. Gee, J.P.: What Video Games Have To Teach Us About Learning And Literacy. Palgrave Macmillan, New York, NY (2007)

4. Oblinger, D.G.: The next generation of educational engagement. J. Interact. Media Educ. **8**(1), 1–18 (2004)
5. Dickey, M.D.: Murder on Grimm Isle: the impact of game narrative design in an educational game-based learning environment. Br. J. Edu. Technol. **42**(3), 456–469 (2011)
6. Hwang, G.J., Chiu, L.Y., Chen, C.H.: A contextual game-based learning approach to improving students' inquiry based learning performance in social studies courses. Comput. Educ. **81**, 13–25 (2015)
7. Kickmeier-Rust, M.D., Albert, D.: Micro-adaptivity: protecting immersion in didactically adaptive digital educational games. J. Comput. Assist. Learn. **26**, 95–105 (2010)
8. Shi, Y.-R., Shih, J.-L.: Game factors and game-based learning design model. Int. J. Comput. Games Technol. (2015). https://doi.org/10.1155/2015/549684
9. Barborsa, A.F.S., Pereira, P.N.M., Dias, J.A.F.F., Silva, F.G.M.: A new methodology of design and development of serious games. Int. J. Comput. Games Technol. (2014). https://doi.org/10.1155/2014/817167
10. Plass, J.L., Homer, B.D., Kinzer, C.K.: Foundations of game-based learning. Educ. Psychol. **50**(4), 258–283 (2016)
11. Kiili, K., de Freitas, S., Arnab, S., Lainema, T.: The design principles for flow experience in educational games. Procedia Comput. Sci. **15**, 78–91 (2012)
12. Ketelhut, D.J., Schifter, C.C.: Teachers and game-based learning: improving understanding of how to increase efficacy of adoption. Comput. Educ. **56**, 539–546 (2011)
13. Sung, H.-Y., Hwang, G.-J.: Facilitating effective digital game based learning behaviors and learning performances of students based on a collaborative knowledge construction strategy. Interact. Learn. Environ. **26**(1), 118–134 (2018). https://doi.org/10.1080/10494820.2017.1283334
14. Brooks, E., et al.: What prevents teachers from using games and gamification tools in Nordic schools? In: Brooks, A.L., Brooks, E., Sylla, C. (eds.) Interactivity, Game Creation, Design, Learning, and Innovation. ArtsIT 2018, DLI 2018. Lecture Notes of the Institute for Computer Sciences, Social Informatics and Telecommunications Engineering, vol. 265, pp. 472–484. Springer, Cham (2018). https://doi.org/10.1007/978-3-030-06134-0_50
15. Squire, K.: Video games in education. Int. J. Intell. Games Simulat. **2**(1), 49–62 (2003)
16. Gros, B.: digital games in education. The design of games-based learning environments. J. Res. Technol. Educ.**40**(1), 23–38 (2007).
17. Gros, B.: Integration of digital games in learning and e-learning environments: connecting experiences and context. In: Lowrie, T., Jorgensen (Zevenbergen), R. (eds.) Digital Games and Mathematics Learning. MEDE, vol. 4, pp. 35–53. Springer, Dordrecht (2015). https://doi.org/10.1007/978-94-017-9517-3_3
18. Papadakis, S.: The use of computer games in classroom environment. Int. J. Teach. Case Stud. **9**(1), (2018). https://doi.org/10.1504/IJTCS.2018.10011113
19. Prensky, M.: Digital natives digital immigrants. Part 1. Horizon **9**(5), 1–6 (2001). https://doi.org/10.1108/10748120110424816
20. Van Eck, R.: Digital game-based learning: it's not just the digital natives who are restless. Educ. Rev. **41**(2), 16–30 (2006)
21. Nousiainen, T., Kangas, M., Rikala, J., Vesisenaho, M.: Teacher competencies in game-based pedagogy. Teach. Teach. Educ. **74**, 85–97 (2018)
22. Plass, J.L., Homer, B.D., Kinzer, C.K.: Foundations of game-based learning. Educ. Psychol. **50**(4), 258–283 (2015)
23. De Grove, F., Bourgonjon, J., Van Looy, J.: Digital games in the classroom? A contextual approach to teachers' adoption intention of digital games in formal education. Comput. Hum. Behav. **28**(6), 2023–2033 (2012)

24. Perini, S., Luglietti, R., Margoudi, M., Oliveira, M., Taisch, M.: Learning and motivational effects of digital game-based learning (DGBL) for manufacturing education – The Life Cycle Assessment (LCA) game. Comput. Ind. **102**, 40–49 (2018)
25. Whang, G.-J., Wu, P.-H.: Advancements and trends in digital game-based learning research: a review of publications in selected journals from 2001 to 2010. Br. J. Edu. Technol. **43**(1), 6–10 (2012)
26. Razak, A.A., Connolly, T.N., Hainey, T.: Teachers' Views on the approach of digital games-based learning within the curriculum for excellence. Int. J. Game-Based Learn. **2**(1), 33–51 (2012)
27. An, Y.-J., Haynes, L., D'Alba, A., Chumney, F.: Using educational computer games in the classroom: science teachers' experiences, attitudes, perceptions, concerns, and support needs. Contemp. Iss. Technol. Teach. Educ. **16**(4), 415–433 (2016)
28. An, Y.: The effects of an online professional development course on teachers' perceptions, attitudes, self-efficacy, and behavioral intentions regarding digital game-based learning. Educ. Tech. Res. Dev. **66**(6), 1505–1527 (2018). https://doi.org/10.1007/s11423-018-9620-z
29. Hébert, C., Jenson, J.: Digital game-based pedagogies: developing teaching strategies for game-based learning. J. Interact. Technol. Pedag. **15** (2019).
30. Hartmann, T., Klimmt, C.: Gender and computer games: exploring females' dislikes. J. Comput.-Mediat. Commun. **11**(4), 910–931 (2006)
31. Bailey, R., Wise, K., Bolls, P.: How avatar customizability affects children's arousal and subjective presence during junk food-sponsored online video games. Cyberpsychol. Behav. **12**(3), 277–283 (2009)
32. Laamarti, F., Eid, M., El Saddik, A.: An overview of serious games. Int. J. Comput. Games Technol. (2014). https://doi.org/10.1155/2020/8953893
33. Hoffman, L.: Learning through games. Commun. ACM **52**(8), 21–22 (2009)
34. Kiili, K.: Digital game-based learning: towards an experiential gaming model. Internet Higher Educ. **8**(1), 13–24 (2005)
35. Huang, W.D., Johnson, T.E., Han, S.-H.: Impact of online instructional game features on college students' perceived motivational support and cognitive investment: a structural equation modeling study. Internet Higher Educ. **17**(1), 58–68 (2013)
36. Barreto, D., Vasconcelos, L., Orey, M.: Motivation and learning engagement through playing math video games. Malaysia J. Learn. Instr. **14**(2), 1–21 (2017)
37. Fairclough, N.: Discourse and Social Change. Polity, Cambridge (1992)
38. Fairclough, N.: Media Discourse, Arnold, Hodder Headline Group, London (1995)
39. Bakhtin, M.: Speech Genres and Other Late Essays. University of Texas Press, Austin (1986)
40. Edwards, D.: Discourse and Cognition. Sage, London (1997)
41. Potter, J.: Representing Reality. Discourse, Rhetoric and Social Construction. Sage, London (1996)
42. Burr, V.: An Introduction to Social Constructionism. Sage, London (1995)
43. Wetherell, M.: Editor's introduction. In: Wetherell, M., Taylor, S., Yates, S.J. (eds.) Discourse Theory and Practice: A Reader, pp. 9–13. Sage, London (2001)
44. Potter, J., Wetherell, M.: Discourse and Social Psychology. Sage, London (1987)

GamAll: Playing Beyond Boundaries - Gamification and Multimodal Literacy

Marcelo Salaberri$^{(\boxtimes)}$ (iD), Maitê Gil, and Cristina Sylla (iD)

Universidade do Minho, 4710-057 Braga, Portugal
{maite.gil,critina.sylla}@ie.uminho.pt

Abstract. Interactive story apps are becoming popular, especially among children. Due to its multimodality, interactive story apps offer a good opportunity to promote the development of cognitive and language skills. Here, we present the theoretical framework and the initial design decisions that support the development of a pedagogical game, which aims at enhancing students' reading competences related to multimodal texts. The game complements Mobeybou in Brazil, an interactive story application directed to pre and primary school children and a digital manipulative that aims at promoting young children's literacy, especially focusing on the development of language and storytelling competences in the context of multiculturalism. The development of the game follows a design-based research methodology and a user-centered approach. The conceptualization and the development of the game, as well as of the Mobeybou story app, are informed by theories of embodiment, socio-constructivist and constructionist theories, as well as multiliteracies and multimodality theories.

Keywords: Gamification · Interactive story apps · Multimodality · Children playful learning

1 Introduction

Mobile devices have infiltrated our daily environment like no other previous technology before [20]. Their learning and interactive functionalities make them popular among young children, who often make their first contact with literature and other subjects through such devices [18]. Although interactive story apps have similarities to printed picture books, the multimodal nature of both is different: while the use of printed picture books predominantly implies a combination of the verbal and the visual modes, story apps usually include auditory, tactile, and performative dimensions [18]. Due to its multimodality, interactive story apps have the potential to offer a unique reading experience, stimulating and conveying information through the visual, auditory and sensory channels [14]. Given that the comprehension of an interactive story app requires specific reading skills and strategies, it is necessary to develop pedagogical materials and tools that support the development of student's reading competences related to multimodal texts.

© ICST Institute for Computer Sciences, Social Informatics and Telecommunications Engineering 2021
Published by Springer Nature Switzerland AG 2021. All Rights Reserved
E. I. Brooks et al. (Eds.): DLI 2020, LNICST 366, pp. 140–147, 2021.
https://doi.org/10.1007/978-3-030-78448-5_10

Taking advantage of the opportunities offered by this new kind of text and the necessary skills to read a multimodal text, this study uses gamification concepts, inserted into a pedagogical game, to scaffold student's reading competences. The game aims at providing opportunities to expand and reinforce the knowledge developed during the reading of the story app in a playful way. In this paper, we present the theoretical framework and the initial design decisions that support the development of the game.

2   Gamification in Educational Contexts

According to Deterding et al., "Gamification is the use of game design elements in non-game contexts" [6, p. 2], even though gamification is sometimes misunderstood as learning through games [2]. At this point, it is important to highlight that this does not mean that learning through games is not possible. In fact, the notion of 'serious games' emphasizes such possibility: "A serious game is a game in which education (in its various forms) is the primary goal, rather than entertainment" [17, p.17]. However, gamification does not imply exclusively a game environment, since its aim is to encourage the user to use the gamified system [16], which can be a game, a task, a questionnaire, and so on. In other words, gamification aims to involve and motivate people through concepts normally found in games, such as feedback, route, score, and competition [7]. The effectiveness of gamification depends on the correct use of its concepts, which can involve participants to an extent that they willingly perform more complex and time-consuming tasks [2, 3]. Furthermore, gamification rather than just the use of game concepts, presupposes the use of game design [5].

2.1   Gamification Concepts

As referred in the previous section a gamified system is composed of concepts that originally belong to games [12]. Based on the orientation proposed by [12], below we list and explain the most frequently used concepts of gamification, that is, points, levels, achievements, feedback, ranking, missions, integration, engagement loops, customization, rules and narrative.

- Points: Are used to quantify the user's actions;
- Levels: Provide the perception of progress in the game;
- Achievements: Are visual representations of the objectives that have been reached;
- Feedback: Is a positive or an indicative reinforcement about the user's performance;
- Ranking: Is used for comparison between people who are participating in the gamified activity, often generating a feeling of competition among them;
- Missions: Are used to guide the user within the system and make it more interesting for her/him;
- Integration: Is the ability of the system to accommodate an inexperienced user or one who uses the system for the first time;
- Engagement Loops: Are techniques to keep the user motivated and willing to use the system again;
- Customization: Allows the user to modify specific items within the game environment according to her/his preferences;

– Rules: Define how the user should use the system, or which permissions s/he has;
– Narrative: Is used to guide the user and encourage him/her to act as expected.

Considering these concepts and guided by our learning goals, we select the concepts that best fit the purpose of our game. The selected concepts go beyond points, ranking and achievements since we use gamification as a learning scaffold and not to promote competition among the users. Therefore, from the concepts presented we chose seven to integrate in our game, namely, levels, points, achievements, feedback, customization, rules and narrative. The way in which these concepts were used will be clarified in Sect. 4.

3 Contextualization

This paper presents the motivation, design and conceptualization of a pedagogical game that complements an interactive story application, Mobeybou in Brazil, which is part of Mobeybou[1], a set of digital and tangible pedagogical materials directed to pre and primary school children [21]. These materials aim at promoting young children's literacy, especially focusing on the development of language and storytelling competences in the context of multiculturalism. The development of the materials and the game follows a design-based research methodology and a user-centered approach [1], involving the target users during several development phases in a cyclical process of designing, testing, redesigning and testing again. The conceptualization and the development of the materials are informed by theories of embodiment [11, 15], socio-constructivist [19] and constructionist [10] theories, as well as multiliteracies and multimodality [8].

3.1 Mobeybou in Brazil

Mobeybou in Brazil presents a story that narrates Iara and Kauê's adventures in Brazil. The application includes a geographical map of Brazil, a 360° environment that replicates a famous street in Brazil, the Avenida Paulista, a puzzle, a page with augmented reality and an incorporated glossary that explains keywords from the story. Children can choose to read the text on their own, listen to the story being read by the default narrator (which also functions as a model reading) or use the integrated recording function to create, record and listen to their own (or a family member or friend's) personalized reading. The recording function supports several audio files, which are then stored in the app. The map helps situate the country in the global world map, as well as the five Brazilian socio-demographic regions. Together with the conveyed knowledge about Brazil, e.g., location, food, traditions, the interactive areas, the 360° environment and the AR experience intend to promote a high level of body expressions and haptic interactions. Another important feature is that the meanings in the interactive story app are represented by different modes (e.g.: texts, images, sounds, animations, possibilities of interaction) having a complementary relationship and creating several layers of interpretation. It is possible to assert that this text is a complex semiotic ensemble, in which different semiotic resources (modes) are combined to create an orchestrated whole [13].

---

[1] http://mobeybou.com/.

Considering multimodality as one of the key meaning-making features in the design of digital texts [4], we argue that the comprehension of an interactive story app requires specific reading skills and strategies. In this sense, the pedagogical game presented in this paper represents one possible answer to this demand.

4 GamAll

4.1 Methodology

As previously referred, this investigation follows a design-based research methodology, encouraging collaboration between participants and researchers [1, 22]. Several methodological procedures are necessary in order to accomplish our main goal, such as: (i) conducting an extensive bibliographic research; (ii) establishing the learning goals; (iii) selecting the gamification concepts that are going to compose the pedagogical game; (iv) detailing how each concept will be explored during the game; (v) conducting the visual and graphic development; (vi) implementing the game (which presupposes a cyclical development focused on designing-testing-improving the game prototype); and (vii) validating the game through empirical investigation. The main learning goal is to scaffold students' reading competences of multimodal texts. Steps (iii) and (iv) are presented in the next section.

4.2 Description of the Game

The choice of gamification concepts to design the pedagogical game was motivated by some of the most cited concepts and the ones that have shown better results to engage and motivate the participants [9], as well as by our learning goals.

Considering the age of the target audience, pre and primary school children, the chosen concepts are easy to understand and do not encourage competition among users, but rather collaboration in order to achieve the goal of expanding and strengthening the knowledge developed through the reading of the story app. The game, still in the development phase, includes the gamification concepts presented below.

– Levels: The game is divided into five regions, (Fig. 1, left), which correspond to the socio demographic regions of Brazil. In each one, there are challenges that address specific characteristics of the region, expanding the information conveyed through the reading of the story app. It is important to highlight that the information about the regions is presented in the story through different modes (visual, aural and haptic), so the game also explores how each mode is understood (or not) by the user. The existence of levels, as presented in the previous section, brings a sense of continuity and progression in the game.

The design for the games in each of the five levels reinforces the multimodal approach. Level 1, for instance, prompts the users to identify objects that are presented mainly through the visual mode in the story (Fig. 2(I)). Level 2 invites the users to discover story elements that were presented through the interplay of verbal and visual modes (Fig. 2(II)); Level 3 challenges the users to pair sound and object, exploring elements presented through aural mode during the story (Fig. 2(III)). Finally, Level 4 asks users to match animals and their habitats, addressing the information presented through the interplay of verbal, visual and aural modes (Fig. 2(IV)). Figure 2 presents the initial design for some of the game level's.

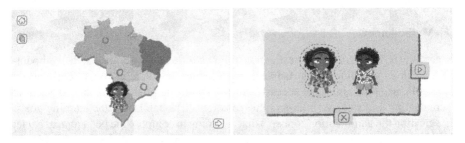

Fig. 1. The map with the levels of the game (left), characters' options (right)

– Customization: The participant will have the opportunity to choose one of two characters to accompany her/him during the trajectory of the game, (Fig. 2, right). S/he will also be able to choose the language (Portuguese/English) of the story, and, finally, the user can decide to listen (or not) to the oral narration. The possibility of customization is a way to help the user to identify with the game.
– Points: Regarding the points, we choose the star format, because it is simple and easy to understand, considering the target audience. The score is divided into three categories:

One star: If the participant does not successfully complete at least 60% of the challenges in a level, s/he will not be able to move to the next phase. So, s/he gets one star and is invited to read the pages about that region in the story app. After reading, the user can try to overcome the challenges again.

Two stars: When the participant gets enough points to advance to the next level, s/he gets two stars.

Three stars: If the user successfully completes more than 80% of the challenges on a level, s/he gets three stars. In this case, in addition to advancing to the next level, s/he receives a reward. The rewards are elements (characters, animals, objects, musical instruments) that are part of other story apps from the Mobeybou set of pedagogical materials.

– Achievements: As explained above, at the end of each level completed with three stars, the user receives a reward. Besides, at the end of the route, if the participant has three stars in all levels, s/he can access and download a file with augmented reality figures from another story app that are part of the Mobeybou pedagogical materials.
– Feedback: Along the game play the user will receive instantaneous feedback and, if necessary, an aid to improve the score, according to the number of stars collected at the

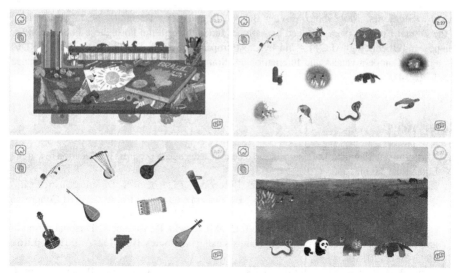

Fig. 2. Initial designs for the some of the game level's: (I) identifying objects related to the story (top-left), (II) discovering the story element (top-right), (III) pairing sound and object (bottom-left), and (IV) matching animals and their habitats (bottom-right)

end of each level. Depending on the users' performance, there will be different kinds of feedback, such as instructions that highlight the semiotic resources used during the meaning-making process, as well as hints that emphasize the relations between each mode, which will be presented during the different challenges in each level. Little tips of extra knowledge will be offered for the users that solve the game. smartphone
- Rules: The rules of the game are defined by the score the participant obtains at each stage, a minimum score is required to proceed to the next stage.
- Narrative: The character chosen in the customization will guide the user throughout the game. Besides, there is a narrative that contextualizes the levels and the challenges the users need to overcome.

5   Conclusion and Future Work

Technologies like smartphones and tablets are already commonly used in educational contexts. Besides, the multimodal nature of digital texts is present in most digital environments. Meanwhile, gamification has proven to be quite powerful as an assistive technique for students, when it comes to motivation and engagement.

In this paper we propose a dialogue between digital environments, multimodality and gamification for the development of a game that aims to reinforce the knowledge constructed through the interaction with an interactive story app for children. Future work includes: the development of various low-fidelity game prototypes that will be tested with the target users, aligned to a design-based methodology; the empirical research of the acceptance of gamification along with multimodality; and, as a final goal, the expansion of the developed game to a framework to guide the development of educational games to support multimodal reading skills.

Acknowledgments. Mobeybou: Moving Beyond Boundaries - Designing Narrative Learning in the Digital Era, has been financed by national funds through the Portuguese Foundation for Science and Technology (FCT) - and by the European Regional Development Fund (ERDF) through the Competitiveness and Internationalisation Operational Program under the reference POCI/01/0145/FEDER/032580.

References

1. Anderson, T., Shattuck, J.: Design-based research: a decade of progress in education research? Educ. Res. **41**(1), 16–25 (2012)
2. Andrade, F.R., Pedro, L.Z., Lopes, A.M.Z., Bittencourt, I.I., Isotani, S.: Desafio do uso de gamificação em sistemas tutores inteligentes baseados em web semântica. In: XXXIII Congresso da Sociedade Brasileira de Computação, vol. 1, pp. 1453–1462 (2013)
3. Challco, G.C., Moreira, D.A., Bittencourt, I.I., Mizoguchi, R., Isotani, S.: Personalization of gamification in collaborative learning contexts using ontologies. IEEE Lat. Am. Trans. **13**(6), 1995–2002 (2015)
4. Cope, B., Kalantzis, M.: "Multiliteracies": new literacies, new learning. Pedagogies Int. J. **4**(3), 164–195 (2009)
5. Deterding, S.: Gamification designing for motivation. Interactions **19**(4), 14–17 (2012)
6. Deterding, S., Dixon, D., Khaled, R., Nacke, L.: From game design elements to gamefulness: defining "gamification". In: Proceedings of the 15th International Academic MindTrek Conference: Envisioning Future Media Environments, pp. 9–15 (2011)
7. Fardo, M.L.: A gamificação aplicada em ambientes de aprendizagem. RENOTE Revista Novas Tecnologias na Educação **11**(1) (2013)
8. Futures, D.S.: A pedagogy of multiliteracies: designing social futures. Harvard Educ. Rev. **66**(1), 60 (1996)
9. Hamari, J., Koivisto, J., Sarsa, H.: Does gamification work? – a literature review of empirical studies on gamification. In: 2014 47th Hawaii International Conference on System Sciences, pp. 3025–3034. IEEE (2014)
10. Harel, I.E., Papert, S.E.: Constructionism. Ablex Publishing, Norwood (1991)
11. Kirsh, D.: Embodied cognition and the magical future of interaction design. ACM Trans. Comput. Hum. Interaction (TOCHI) **20**(1), 1–30 (2013). https://doi.org/10.1145/2442106.2442109
12. Klock, A.C.T., de Carvalho, M.F., Rosa, B.E., Gasparini, I.: Análise das técnicas de gamificação em ambientes virtuais de aprendizagem. RENOTE-Revista Novas Tecnologias na Educação **12**(2) (2014)
13. Kress, G.R.: Multimodality: A Social Semiotic Approach to Contemporary Communication. Taylor & Francis, New York (2010)
14. Kucirkova, N.: Theorising materiality in children's digital books. Libri Liberi **8**(2), 279–292 (2019). https://doi.org/10.21066/carcl.libri.8.2.2
15. Lakoff, G., Johnson, M.: Philosophy in the Flesh: The Embodied Mind and its Challenge to Western Thought, vol. 640. Basic books, New York (1999)
16. Landers, R.N.: Developing a theory of gamified learning: Linking serious games and gamification of learning. Simul. Gaming **45**(6), 752–768 (2014). https://doi.org/10.1177/1046878114563660
17. Michael, D.R., Chen, S.L.: Serious games: Games that educate, train, and inform. Muska & Lipman/Premier-Trade (2005)
18. Nikolajeva, M., Al-Yaqout, G.: Re-conceptualising picturebook theory in the digital age. Nordic J. Childlit Aesthetics **6** (2015). https://doi.org/10.3402/blft.v6.26971

19. Piaget, J.: The development of thought: Equilibration of cognitive structures. (Trans A. Rosin). Viking (1977)
20. Read, J., Markopoulos, P.: Int. J. Child-Comput. Interaction. https://doi.org/10.1016/j.ijcci.2012.09.001
21. Sylla, C., Pires Pereira, Í.S., Sá, G.: Designing manipulative tools for creative multi and cross-cultural storytelling. In: Proceedings of the 2019 on Creativity and Cognition, pp. 396–406 (2019). https://doi.org/10.1145/3325480.3325501
22. Wang, F., Hannafin, M.J.: Design-based research and technology-enhanced learning environments. Educ. Tech. Res. Dev. **53**(4), 5–23 (2005)

Designing a Learning Robot to Encourage Collaboration Between Children

Wouter Kaag(✉), Mariët Theune, and Theo Huibers

University of Twente, Drienerlolaan 5, 7522 NB Enschede, The Netherlands
k.w.kaag@alumnus.utwente.nl,
{m.theune,t.w.c.huibers}@utwente.nl

Abstract. Collaboration is an important skill for children to learn. In this paper we present a small-scale study exploring how technology can be used to elicit collaboration between children. We developed a prototype of a tablet-based robot ('surfacebot') that tried to perform a specific task, while children acted as tutors by giving feedback on the surfacebot's actions. The surfacebot used the feedback to improve its actions by means of reinforcement learning. A pilot study with an early prototype showed that children were engaged and provided consistent feedback to the surfacebot, but showed little collaboration. Instead they made individual decisions and took turns in providing feedback. Based on these observations we made several changes to the prototype that were meant to stimulate collaboration between the children. In our main study with the revised prototype, we measured collaboration using an annotation scheme based on a collaborative problem solving framework. The results suggest that the revisions of the prototype indeed led to more extensive collaborative behavior, with 4 of the 9 participating pairs of children establishing a division of roles that necessitated perspective taking and mutual exchange of information.

Keywords: Collaboration · Primary school children · Learning robot

1 Introduction

Collaboration, referred to as a 21st century skill [1,5], is an important skill for children to learn since it relates to critical thinking, meta-cognition and motivation [14]. Improving a skill requires practice, and there might be a long-term education gain when children discover collaboration for themselves [2]. It has been shown that beneficial effects regarding learning and development, particularly in the early years or primary education, can occur when children work in pairs or small groups [25]. Furthermore, self-esteem and attitudes towards others are mentioned as beneficial outcomes of collaborative learning in the classroom [4,21]. But collaboration can be challenging for young primary school children, as children below the age of 7 have not yet developed all of the cognitive skills

© ICST Institute for Computer Sciences, Social Informatics and Telecommunications Engineering 2021
Published by Springer Nature Switzerland AG 2021. All Rights Reserved
E. I. Brooks et al. (Eds.): DLI 2020, LNICST 366, pp. 148–168, 2021.
https://doi.org/10.1007/978-3-030-78448-5_11

required for effective peer collaboration, such as recursive perspective taking [26]. Therefore, this study focused on using technology to foster collaboration between children.

To encourage collaboration between children, we devised an interactive technology concept with a robot as a teachable agent. Computer agents taught by children are known as teachable agents [3]. As our teachable agent we used a surfacebot, see Fig. 1. Surfacebots were originally developed as affordable, mobile and flexible robots to be used in collaborative storytelling activities on a non-digital tabletop [6]. A surfacebot has two parts: a tablet and a base with wheels, making it capable of movement, sound and visual representations. The tablet can be used as a character display [24] and as an interactive interface [7].

Fig. 1. The surfacebot as used in our study.

Our objective was to explore how an activity with a surfacebot can be designed to encourage collaboration between pairs of primary school children. We expected that primary school children could benefit from settings that encourage social interactions and collaboration, in particular children in the age of 5–7, because their collaboration skills start developing [2]. Our main research question was: *How can the capabilities of the surfacebot be utilized to create an engaging activity that effectively encourages collaboration between children?* In order to answer this question, we iteratively designed prototypes and tested them with children. We also developed an annotation scheme based on the framework for collaborative problem solving skills by Hesse et al. [11] for evaluating the level of collaboration between children during an activity.

Firstly, in Sect. 2 we describe several applications designed for collaborative activities. Section 3 describes our first concept to stimulate collaboration between children using the surfacebot. In the pilot study (Sect. 4), the degree of collaboration between children using this prototype is explored. The findings from the pilot study lead to a second version of the prototype (Sect. 5). In the main study (Sect. 6), we analyzed the level of collaboration using this new prototype. We end with a discussion of the results (Sect. 7) and our conclusions and recommendations (Sect. 8).

2   Related Work

Much work on collaborative technologies for children has been carried out in the context of creative applications such as collaborative storytelling for children [10]. Early examples are KidPad and the Klump: storytelling technologies that allowed children to work independently, but encouraged collaboration by providing added benefits of collaborative actions in terms of efficiency or fun [2]. A more extreme approach to collaboration was taken in the Story Table system, which forced children to work together by requiring multiple-user touch actions (performed by children simultaneously) for certain crucial operations [28]. In a non-storytelling context, Woodward et al. tried to encourage collaboration through role division in a digital tabletop game for children [27].

Some work has been done on collaborative storytelling for children with robots moving on a tabletop. An example is RoboTale, of which the main character is a robot similar to the surfacebot. RoboTale successfully stimulated collaboration among small groups of school children in the form of passing tangibles to each other (prompted by the size of the table and scarcity of resources) and discussing the plot of the story they created together [15]. Earlier work using the surfacebot as a teachable agent had it as the main character in a playful story-based activity, traveling around 'in France' (on the tabletop) and being taught French words by pairs of children [24]. In this activity, collaboration was enforced through a fixed role division, where one child was teaching the words to the surfacebot, and the other child was moving the surfacebot.

Other research on robots as teachable agents explored how children perceive and correct the handwriting of a robot. A study with 24 children (aged 7–8) acting individually as handwriting tutors of a Nao robot showed that these children paid attention to the learning of a robot and were capable of providing corrections using a slider or by demonstration [8]. For a survey of studies with robots as teachable agents for children, see [13].

A major inspiration for our work was Sophie's Kitchen [22], an application featuring a virtual agent (Sophie) that learned from human feedback how to bake a cake, using a reinforcement learning algorithm called Q-learning. Experiments were carried out with different versions of Sophie's Kitchen, to investigate how adult users wanted to teach the agent. The first experiment allowed users to provide feedback using a slider. The results showed that being able to guide the agent's attention through feedback resulted in a faster learning interaction, compared to only providing feedback after the fact. In another experiment Sophie used gazing behavior to indicate which action she was about to take ('transparency behavior'). This led to people providing guidance more often when it was required and less often when not. Another version of Sophie had so-called 'undo' behavior: retracting an action (if possible) after receiving negative feedback. This was shown to further improve the learning behavior of the agent.

Our work combines various aspects of these prior works, while also showing some important differences. Our teachable agent is like Sophie in that it uses Q-learning to learn from human feedback, but our users are pairs of children instead of individual adults. Like [8,24] we use a robot as our teachable agent, but our primary aim is not learning-by-teaching but stimulating collaboration between children through the teaching activity. Finally, like [2,15] we try to nudge children into collaboration through the design of the activity, instead of forcing it like [28].

3   First Prototype

Collaboration involves a "mutual engagement of participants in a coordinated effort to solve a problem together" [20, p. 70] and is affected by the structure and design of a task [14]. We kept this in mind as we developed a first prototype of a collaborative activity with the surfacebot as a teachable agent. To encourage children to collaborate while teaching the surfacebot, we created a story that portrays the surfacebot as a bear called Ted (shown on the surfacebot's tablet screen) who wants to get dressed to go outside, but needs the help of the children to find the right clothes as he does not know which clothes fit the weather.

The clothing items are spread across four locations on the tabletop, each representing a different room in the bear's house and showing several associated items of clothing (printed on cards) associated to that location, see Fig. 2. The surfacebot moves around these locations and selects clothing items to wear. Children can provide feedback on the surfacebot's actions using a slider. The idea is that the surfacebot learns from the feedback and adjusts its decision making based on it. At some point, it decides to 'go outside' indicating the end of a round. If the selected clothes do not match the weather, another round starts. These actions and learning ability of the surfacebot were simulated in the first prototype.

Collaboration is characterized by a symmetrical structure: a symmetry of goals, actions, knowledge and status [9]. Therefore the activity was designed to have a symmetrical structure to encourage collaboration. This was done by giving children a shared goal of assisting the surfacebot in completing its task, as well as equal knowledge and opportunities to interact with the surfacebot.

The prototype consisted of three parts, see Fig. 3. The first part is an application for the surfacebot's tablet, further referred to as the 'character display'. It displays a bear, the clothes it is wearing and the thoughts it has; see Figs. 2 and 3 (middle). Preceding an action, the bear would think about a piece of clothing, shown in a thought cloud, see Fig. 4(a). After 3 s, the thought disappeared and the bear could be seen wearing the item, see Fig. 4(b). The thought cloud was included as a 'transparency behavior' [22] of the surfacebot about its upcoming actions (clothes choices). It was used to give children the time to act and to pave the way for eventually implementing a form of undo behavior, which could lead to faster learning of the surfacebot [22].

The second part of the prototype was the 'reward interface', through which children could communicate feedback to the surfacebot; see Fig. 3 (left). It included two interactive elements: a slider and a 'send' button (see Fig. 6(a) for a more close-up view). The slider enabled children to communicate a degree of right or wrong. The idea was that the slider could stimulate negotiation, and

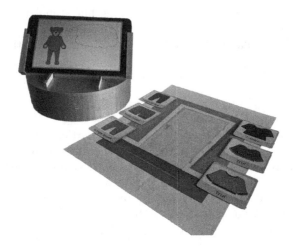

Fig. 2. Impression of the first prototype. The surfacebot with the character display is positioned next to one of the four locations with its associated clothes cards.

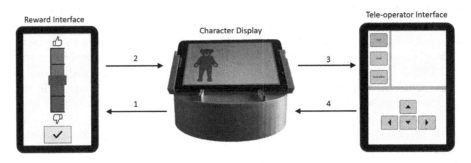

Fig. 3. The three main components of the prototype. The character display is the server, and communicates with the clients: the reward (left) and tele-operator (right) interfaces. The communication involves (1) status of the activity, (2) value and timing of feedback, (3) the name of next action according to the script and (4) controlling the activity (e.g. starting the next action) and controlling the surfacebot's movement.

(a) Thought   (b) Action   (c) Notification

Fig. 4. Character display showing (a) what clothing item the bear is thinking of, (b) the bear wearing the item, and (c) a feedback notification (thumbs up or down).

could be used to reach a consensus when opinions differed, for example by going for an intermediate value. The send button was used to confirm the feedback. When children communicated feedback, a notification appeared on the character display, see Fig. 4(c).

The third part is the tele-operator interface; see Fig. 3 (right). In the first prototype, the surfacebot had no learning ability yet and could not move autonomously. In our tests with this prototype, a Wizard-of-Oz approach was used to make it seem like the robot acted and moved autonomously and that their feedback had an effect on the robot's choices. The tele-operator interface was used to control the activity and fake the autonomous behavior of the surfacebot, following a script with sequences of actions to simulate learning. Over the course of three rounds, the surfacebot's actions became increasingly more accurate and ultimately led to a set of clothes appropriate to the weather scenario.

4   Pilot Study

The first prototype was tested in a pilot study with 12 children at a local daycare facility. The study was approved by the Ethical Committee of our faculty and the parents of the children had given consent for their participation. The age of the children varied between 4 and 8 years old, with an average age of 5.75 years. With the study, we aimed to validate the concept and identify areas of improvement by getting a first impression of how children engaged, collaborated and provided feedback in the designed activity.

4.1   Method

In the pilot study, 6 pairs of children participated in sessions of ±10 min. The tests were conducted in a separate room at the daycare facility. Figure 5 gives a top view impression of the setup. The procedure was as follows. First, the children received a short introduction in which the setup, the task and how the tablet worked was explained and demonstrated. Then they began their first round, giving feedback to the surfacebot as it was putting on clothes, until the surfacebot decided to 'go outside'. During the activity, the facilitator was always present to answer the children's questions. The children were not guided or motivated by the facilitator to provide feedback and could stop the activity at any time they liked. After three rounds, the children were thanked for their participation and helpfulness. They were made aware of the Wizard-of-Oz method (the facilitator controlling the surfacebot) used during the activity.

How the pairs of children acted and collaborated in the activity was observed during the test, and notes were taken afterwards. The sessions were recorded if consent for this was given by children's guardians. Tablet interactions were logged on the tablet to get an insight into the frequency and values of the children's feedback.

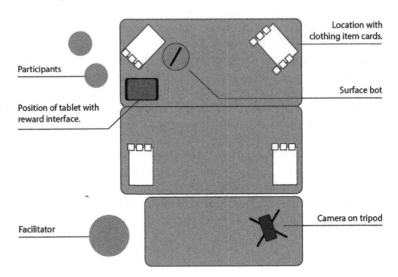

Fig. 5. Impression of the setup of the pilot study. The participants started the activity near the reward interface. The surfacebot moved between the locations at the corners of the table. The camera, on a tripod, was located on a table next to the activity. The facilitator also remained nearby for controlling the surfacebot.

4.2 Results

Limited collaboration between children was observed. Collaboration mostly took the form of providing feedback to the robot in turns. There was little conflict or negotiation about the task, but rather about who operated the tablet. Most children decided to walk around with the tablet, which created a situation where the activity was easily doable for one child. It allowed one child to track the robot's actions and communicate feedback at the same time. This may have provided little incentive for collaboration. This is confirmed by one pair of children who were observed to divide roles. During their session, the tablet remained on the table. It resulted in a situation where the children relied on each other to share information and opinions about the robot's actions. One followed and observed the robot while the other stayed close to the tablet and communicated feedback.

Children were observed to be motivated and enthusiastic during the activity. The logged data showed that the majority of the children continued to be engaged by consistently providing feedback, except for one pair who were both very young and shy (age 4 and 5). They had little communication and did not make use of the tablet. The logged data also showed that children used the reward interface in a binary way, with mainly extreme feedback values being communicated to the robot. It might be due to a certain unanimity that children had about which actions of the surfacebot were right and wrong. This made the slider somewhat superfluous.

The results correspond to the characteristics of the developmental stage of children aged 3 to 7 years, who are self-centered and prefer parallel play [16].

The majority of the children were engaged in the activity and provided feedback throughout the activity. This suggests that children could handle their role as tutor, and remained motivated during the activity. We concluded that ensuring a symmetrical structure [9] allows collaboration, but is not enough to elicit it.

5   Second Prototype

We modified the first prototype based on the insights obtained from the pilot in order to encourage collaboration more effectively.

Fixed Tablet. Firstly, an important change to the setup was made by fixing the tablet to the table. This would make observing the robot and giving feedback at the same time more difficult, so we expected this to stimulate collaboration in the form of a role division [27].

Improved Reward Interface. A new reward interface was created, see Fig. 6(b). The green-red gradient represents the transition from completely right to completely wrong. The aim was to clarify the use of the slider to encourage more diverse feedback.

Ambiguity. Ambiguous tasks tend to foster collaboration, as disagreements and misunderstandings can cause communication in the form of explanations and reasons [9]. Therefore, more ambiguity was added to the activity by providing more choices between similar items of clothing. Also, the names of the items shown on the cards were made more general to ensure they did not hint to a certain outfit or type of weather. For example, one item used in the first prototype had the description: "winter shoes". This was changed to: "shoes". It was expected that the less specific names could lead to more discussion among the children about the objects, which would require collaboration to provide unanimous feedback.

Reinforcement Learning. The prototype was extended with actual autonomous behavior of the robot, making the robot capable of choosing its own actions. The role of the tele-operator was limited to navigating the surfacebot to its next location (picked by the surfacebot). The script used for the robot's actions was replaced by a Q-learning framework, inspired by the work of [22]. It enabled the surfacebot to take actions and to learn from feedback. Q-learning is a form of reinforcement learning where an agent derives the optimal policy on how to act in the current state of its environment, given a transition model that describes the state transitions –the state resulting from an action in a given state– and a reward function which contains the reward received based on a state transition. In our prototype the surfacebot's state was based on the clothes it was wearing. The children's feedback determined the rewards on the robot's actions to maximize the cumulative reward, meaning: which clothes lead to most positive feedback. The surfacebot decided to either explore new states or exploit known states. Exploration means taking random actions in order to get information about the reward of being in a certain state. Exploitation means going for the action with the highest reward. As action selection strategy, the surfacebot used the epsilon-greedy [23] approach. This strategy enabled an increasing focus on exploiting what was learned, which led to the surfacebot picking the right clothes faster in the later rounds of the activity given that children provided feedback regularly. In other words: at first the robot would try on clothes at random, collecting feedback on its choices, and gradually it would focus more on picking clothes for which it had received positive feedback earlier.

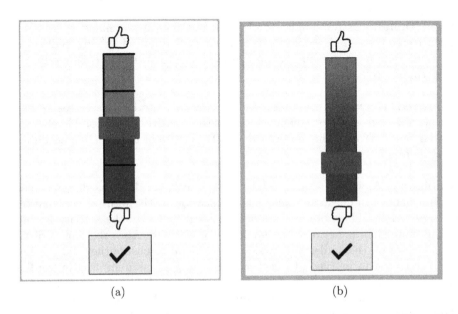

Fig. 6. The reward interface used in the first prototype (a), and the improved interface (b) used in the second prototype. The slider can be used to determine the feedback value. The send button below the slider sends the feedback to the surfacebot.

Undo Behavior. Also inspired by [22], an action cancelling behavior was implemented to provide an immediate response by the surfacebot. If negative feedback was received while the surfacebot was 'thinking' about a certain action, the action would be cancelled and the thought cloud would show a different action.

6   Main Study

In the pilot children were engaged and provided frequent feedback, but showed limited collaboration in the activity with the first prototype. In the main study, we explored to which extent the improvements made to the prototype encouraged collaboration. To analyse the collaborative behavior, an annotation scheme was developed for assessing the level of collaboration between children.

6.1   Participants

The study was conducted with 9 pairs of children at a primary school. The age of the children was between 6 and 10 years (mean = 8.00). Two different classes were involved, but it was ensured that the pairs consisted of children from the same class. The parents or guardians of the participating children received a brochure informing them of the nature of the study and the data being collected, and gave consent for their child's participation and for video recording the sessions.

6.2   Setup

The tests were conducted on two different days. The first day, pairs 1–6 participated in the study. One week later, pairs 7–9 participated. The tests were held in two different classrooms, but both rooms were considered to be a familiar setting to the children. The setup was the same as that of the pilot study.

6.3   Procedure

The procedure used in this study was largely similar to the pilot study. The main difference was that before the start of the activity, we tried to get insight into the level of agreement in the preferences of the children regarding the clothing items. The facilitator told the story introducing the surfacebot, and explained the setup and activity. Then the children were shown a list of all clothing items and had to indicate individually which of the items they preferred. At this point, they had not had the opportunity yet to share or discuss their preferences. We expected that if the children's preferences differed, they would need to negotiate to provide consistent feedback to the robot, therefore displaying a higher degree of collaboration. If children were completely in agreement on the items of clothing the robot should get, the task would become unambiguous and there would be little need for negotiation.

6.4   Annotation Scheme for Assessing Collaboration

We developed an annotation scheme to assess the level of collaboration between children during the activity with the surfacebot. The scheme was based on the collaborative problem solving framework of Hesse et al. [11]. Their framework consists of indicators of social skills that form the 'collaborative' part and indicators of cognitive skills that constitute the 'problem solving' part. Our focus was on evaluating the level of collaboration between children, not their skill in solving the task, therefore we used only the indicators of the collaborative part.

Hesse et al. distinguish three classes of indicators that subsume the social skills: participation, perspective taking and social regulation [11]. We took their definitions of the indicators as a starting point and adapted them to assess the level of collaboration between two children in the context of our task, see Table 1. Below we explain how we used the indicators, supported by invented dialogue examples inspired by the observed behaviour of pairs in the pilot study.

Table 1. The annotation scheme used in the main study. Each indicator is listed followed by the definition of Hesse et al. [11], and our interpretation of it in this study.

Indicator	Definition	Description
Participation		
Action	Acts within environment	A child operates the tablet e.g., sends a reward
Interaction	Interacts with, prompts and responds to the contribution of others	A child interacts with the other child
Perspective taking		
Adaptive responsiveness	Ignores, accepts or adapts contributions of others	A child responds to the contribution of another child e.g., disagreeing with a reward sent
Audience awareness	Awareness of how to adapt behaviour to increase suitability for others	A child shares information that is not available to the other e.g., what action the robot thinks of
Social regulation		
Negotiation	Achieves a resolution or reaches compromise	A child attempts to reach an understanding with the other child e.g., about the value of the reward
Self evaluation	Recognises own strengths and weaknesses	A child vocally reflects on an (inter) action regarding the activity
Trans-active memory	Recognises strengths and weaknesses of others	A child vocally reflects on an (inter) action of the other child
Responsibility initiative	Assumes responsibility for ensuring parts of task are completed by the group	A child makes an effort to involve the other child in the activity

Participation. This class consists of the indicators *action* and *interaction*. *Action* was described as "participation of an individual, irrespective of whether this action is in any way coordinated with the efforts of other group members" [11, p.42]. In the context of our activity, we defined actions as the possible interactions with the robot, i.e. operating the reward interface. *Interaction* takes place when children respond to a contribution of another child, i.e. a comment, question or action. The action and interaction indicators served to provide us with an estimation of the level of the children's engagement with the activity and with each other during the experiment. A third indicator for participation in the framework is *task completion* [11]. This indicator was not applicable in our case, since children were only tasked to provide feedback and the robot decided when the task was completed.

Perspective Taking. The indicators of this class are *adaptive responsiveness* and *audience awareness*. It is considered adaptive responsiveness when a child accepts or adapts another child's point of view. We defined *audience awareness* as a child sharing information that was not available to the other child. For example, if one child is the only one able to see the surfacebot's screen, and tells the other child what action the bear wants to take, this shows that the child is aware of the other's perspective, and shares the information accordingly.

Social Regulation. This class has as its indicators *negotiation, self-evaluation, trans-active memory* and *responsibility initiative*. We defined negotiation as an attempt by the children to reach a common understanding, achieve a solution, or reach a compromise. An example of how children could negotiate using the prototype:

> *Child A: "Item x is super wrong", and sets slider to absolute negative.*
>
> *Child B: "No, it is a bit wrong, but not super wrong."*
>
> *Child A: Adjusts the feedback slider in accordance to feedback of Child B.*

Self-evaluation concerns any comments of a child on their own performance in terms of appropriateness or adequacy in context of (inter)actions during the activity. It indicates a child's recognition of their own strengths and weaknesses. To illustrate, a child could say: "I was too late, now the robot wears the wrong jacket." Conversely, trans-active memory is when one of the children comments on the performance of the other in terms of appropriateness or adequacy. For example: "Let me operate the tablet, you were not fast enough!" Lastly, responsibility initiative is about involving others in the task of learning the robot. An indication is the use of first-person plural in communication regarding the activity, for example: "We should let the bear know that it is the wrong item!" In this study, taking responsibility was also annotated when one child encouraged the other to take action or share information. An example of responsibility initiative:

> *Child A: "Item x is OK, right?"*
>
> *Child B: "Yes, send the feedback!"*

6.5   Identifying Collaborative Behaviour

The annotation scheme described above mostly focuses on the analysis of children's individual utterances. However, we felt that collaboration can also take the form of certain types of more overarching joint behaviour between children.

First of all, we considered a *division of roles* to be a form of collaboration. It happens when children divide responsibilities and a form of interdependence arises in order to successfully complete the task. An example of a division of roles we observed in the pilot study is that one child operated the tablet while the other child tracked the robot and provided updates on the robot's actions.

Another form of collaboration is *shared planning*. A shared planning is established when the problem at hand is analyzed and a mutual agreement is reached on how to approach it. A case of shared planning in the context of the activity would be: an agreement between children about how an action of the surfacebot should be judged, as in the following example:

Child A: *"The bear should get this jacket... and then go to the hallway."*

Child B: *"No, it should get the sweater first then go to the hallway."*

Child A: *"OK, jacket, sweater and then it should get these shoes in the hallway."*

Besides a shared planning, children can also build *shared knowledge* by getting an understanding of another child's opinion or preferences, or establishing a shared understanding of (an element in) the activity. It applies when a child provides information or shares an opinion, in response to a question or statement of the other child. An example of shared knowledge is:

Child A: *"What items do you think the bear should get?"*

Child B: *"The jacket and the blue jeans!"*

Or when roles have been divided:

Child A: *"What is the item that the robot displays?"*

Child B: *"It is the red jacket."*

Lastly, we considered whether there was any *turn taking* between the children. The pilot study showed that children occasionally took turns in operating the tablet, or even played the game individually. Taking turns can only occur after establishing an agreement and can therefore be seen as a result of collaboration. However, it is not necessarily a wanted outcome. In the pilot study, turn taking led to situations where one child temporarily did not actively participate in the activity while the other did everything. It means they did not discover the benefit of collaboration, but made a kind of compromise to each be in full control of the activity for a while. On the other hand, taking turns could also occur while maintaining a division of roles. For example, one operates the tablet, while the other follows the surfacebot from location to location. After a while, they might decide to switch roles. This would be a coordinated effort that maintains the situation where children depend on each other while they both actively participate.

6.6 Measurements

Results were obtained via (1) observations of the video recordings, annotated using the annotation scheme, (2) logged data from interactions with the reward interface, and (3) the clothing preferences that were filled in individually by each participant before the start of the activity.

The video recordings were used for assessing the level of collaboration using the annotation scheme. They were annotated from the point the robot started taking action until the last iteration, where the robot went 'outside'. Our annotation method was inspired by the work of Huskens et al. [12], who evaluated children's collaborative play by reviewing 10 s fragments of videotaped play sessions for a fixed set of behaviors. A behavior that was present in the fragment was recorded as a plus. A behavior that was absent was recorded as a minus.

We adopted a similar approach by reviewing 30 s intervals of the recorded sessions using the annotation scheme. Since we wanted to observe the collaboration between children, we opted for a broader interval than 10 s. Each interval was annotated for the presence of any of the collaboration indicators described above. When an indicator was observed, it was marked as positive $(+)$, otherwise as negative $(-)$.

Subsequently, we determined per session a score by summing the positive annotations of the intervals for each category of the annotation scheme. This resulted in 8 indicator scores of which we took the average as an overall collaboration score. Since the sessions differed in number of annotated intervals, we computed both the indicator score (Eq. 1) and the collaboration score (Eq. 2) proportionally to the total number of intervals.

$$indicator\ score = \frac{\sum annotated_{positive}}{n_{intervals}} \tag{1}$$

$$collaboration\ score = \frac{\sum indicator\ score}{n_{indicator\ scores}} \tag{2}$$

In the same way, we calculated class (e.g. perspective taking) scores by averaging over the scores of the associated indicators. We then used these scores to compare the level of collaboration between pairs of children.

To check the reliability of the annotation scheme, one recording was annotated by two of the authors. The annotations of both authors were almost the same, except for a few minor differences due to differing interpretations of some indicators. Based on this, the definitions of each indicator were refined and example dialogues were added. This resulted in the indicators described in the previous section. Which general forms of collaborative behaviour (e.g., turn taking) took place was determined per recording, instead of per interval.

6.7 Results

The annotations of the video recordings showed that four pairs of children established a role division, where one operated the tablet and the other tracked the surfacebot. In this respect, the second prototype encouraged collaboration more effectively, compared to the single observation of a role division during the pilot. The biggest trigger of a division of roles was the fixed tablet, which made following the surfacebot and giving feedback at the same time difficult for one individual, providing an incentive to divide roles. Figure 7 gives an impression of the children's positioning at the start of the activity and when roles were divided. One child communicates feedback to the surfacebot based on the input of the child that follows the surfacebot, who either shares information about the actions of the robot or shares his/her opinion about the action of the robot. A shared planning was observed for five pairs. Building shared knowledge occurred for three pairs. Children taking turns happened for six of the nine pairs. Each pair that established a role division also took turns. The role of operating the tablet seemed favorable to the children, so they switched roles occasionally.

Fig. 7. The image on the left shows an impression of children at the start of the activity. The image on the right shows that they have established a division of roles.

We annotated the recordings of the 9 sessions using the scheme described in Sect. 6.4, with an average of 17.78 ± 3.55 intervals of 30 s per session. For each pair of children, we calculated the average score for each indicator, as well as average scores for the three classes of indicators: participation, perspective taking and social regulation. We also calculated an overall collaboration score based on the average indicator scores.

Because we suspected that the presence or absence of a role division might have an effect on other collaboration aspects, we compared indicator and class scores of pairs based on whether the children had established a role division or not, see Fig. 8. Across the indicators, the pairs that had a division of roles scored higher on other aspects of collaboration as well. This is reflected in the average scores of the three classes and the overall collaboration score, see Fig. 9. Most notably, the four pairs with an observed role division had a considerably larger score for audience awareness compared to the five pairs that did not. The way audience awareness was annotated was to a large extent aligned with how children divided roles. Each role division resulted in one child describing the surfacebot's action in situations when the other could not see the character display, which was annotated as a form of audience awareness.

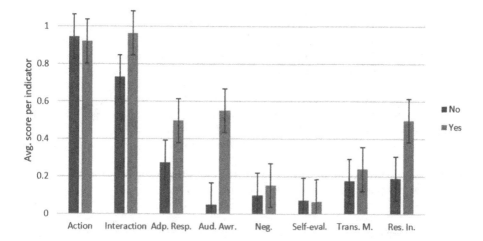

Fig. 8. The average collaboration indicator scores for pairs that established a role division (yes, n = 4) compared to the pairs that did not (no, n = 5).

Two pairs had a very low total collaboration score, 0.14 and 0.25, compared to scores between .38 and .54 for the other pairs. These two pairs communicated less regarding the surfacebot's actions in comparison to the other groups, and did not establish a division of roles. Remarkable is that all four children in these two pairs were 9 years old, and the oldest among the participating children. Perhaps they found the activity too simple or childish, which resulted in less commitment and engagement compared to the younger children. It could indicate that the current prototype is indeed most suitable for children aged 5 to 7 years.

The use of the slider remained unchanged compared to the pilot study, in spite of the ambiguity introduced in the task and the redesign of the slider. Given the value range of 0 to 1 for the feedback, 78.7% of the communicated feedback was an extreme value, either between 0 and 0.1 (37.7%) or between 0.9 and 1.0 (41%). Therefore, providing an incentive to negotiate by leaving room for disagreement did not lead to a more sophisticated use of the slider. However, forms of negotiation were observed where children tried to determine the slider value together. In that sense, the slider still had added value.

We examined to which extent the children's initial clothing preferences overlapped. None of the pairs had exactly the same preferences, which suggests there was room for discussion and negotiation. However, we found no indication that children who had little agreement beforehand showed more collaboration, neither in the overall collaboration score, nor in the 'negotiation' indicator.

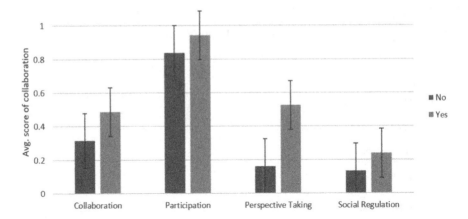

Fig. 9. The average scores of each class for the pairs that established a role division (yes, n = 4) compared to the pairs that did not (no, n = 5).

7   Discussion

Judging from the results of the main study, the second prototype seemed to encourage collaboration more effectively compared to the first prototype. Analysis of the recordings showed a higher average collaboration score for the pairs of children who established a role division, compared to the average score of the couples who did not. This difference is mainly due to higher scores for the 'audience awareness' and 'responsibility initiative' indicators. These indicators relate to sharing new information and encouraging others to take an action, which manifested itself when children divided roles.

The collaboration scores of the pairs of children appear to be mostly influenced by the presence of a role division, characterised by interdependence and communication between children. Interdependence is an influencing factor of collaboration [14]. Communication is described as an integral element of collaboration [11] and an interpersonal skill that will develop when children are provided with the opportunities for social interaction [6]. Rather than explicitly assigning roles to the children as in [27], we left it up to the children whether they established a role division or not, and in what way, thus allowing them to adopt their own collaboration style. Supporting diversity in group dynamics has been argued to be beneficial to collaborative learning [17].

However, there are other factors that may have played a role. The children were in a phase where social skills develop, and age may have influenced the level of collaboration. Another factor that we did not take into account was the level of closeness between the pairs of children, who might have been friends or just classmates. It is therefore difficult to attribute any differences in collaboration purely to changes made to the prototype or to an established division of roles.

Unlike earlier studies of collaboration between children [19, 27], which were purely qualitative in nature, we attempted to carry out a quantitative analysis based on observations. The validity of the scoring system we used still requires further investigation. An overall score for collaboration was calculated by weighting each indicator equally. However, it can be argued that some indicators are more indicative or more important for measuring collaboration than others. For example, 'action' only says something about the participation of children while 'negotiation' is a profound expression of collaboration that requires communication and a certain willingness of both children to listen to each other. In future research, a method could be developed to achieve a more sophisticated collaboration score. The indicators could, for example, be weighted according to their importance or contribution to measuring collaboration.

Children aged 3 to 7 are in a developmental phase where they enjoy fantasy [16], so an activity with a robotic character and a story can be appealing to them. However, the second prototype seems to be mainly suited for children in the age 5–7, since in the pilot study children younger than 5 showed limited understanding of the activity, while in the main study children aged 9 had much lower collaboration scores and seemed less engaged in the activity. In order to reliably compare new versions of the prototype regarding the level of collaboration displayed by children, it would be better to keep the age difference between the (pairs of) children as small as possible.

Finally, due to the small numbers of participants we cannot draw any strong conclusions at this stage; studies with more pairs of children are needed to confirm our preliminary conclusions.

8   Conclusion and Future Work

We iteratively developed a prototype of the surfacebot as a teachable robot designed to encourage collaboration. A pilot study showed that our first prototype only brought about a limited form of collaboration, as most children took turns with only one child actively teaching the robot. In the main study with a revised prototype, multiple pairs of children established collaboration through a spontaneous division of roles, with one of them operating the tablet and the other providing information about the robot's actions or sharing opinions about them.

Although children showed enthusiasm and established collaboration while interacting with the prototype, it was outside the scope of this research to determine if children learned or developed their collaboration skills from participating in the activity. A recommendation for further research is to conduct a study into the long term effect of participating in activities with the surfacebot (that are aimed to encourage collaboration) on the collaborative skills of primary school children. Do children who regularly participate in such an activity with the surfacebot show improved collaboration compared to children who did not use the surfacebot? And if children improve their collaborative skills through the activities with the surfacebot, are the knowledge and skills transferred to other collaborative activities without the surfacebot? Longitudinal studies could reveal the educative contribution of activities with the surfacebot when introduced in the classroom.

Secondly, we recommend to redesign the activity towards a more ambiguous and complex task that invites using the slider more and gives rise to more discussion. In the current activity, children showed an understanding of the interface and the slider's purpose and effect. However, they mainly provided unilateral feedback with only completely negative or completely positive values. There was low uncertainty or disagreement about the correctness of the surfacebot's actions, limiting the need for negotiation.

Finally, we recommend exploring how learning material from children's regular school subjects can be incorporated into the concept in order to optimally exploit the learning-by-teaching paradigm, while maintaining an activity design that encourages collaboration. Curriculum-focused design techniques could be used to determine which educational topics and activities could benefit from our technology [18].

Acknowledgements. We thank Alejandro Catalá Bolós for the use of the surfacebot. We also thank the children who participated in our experiments, the daycare and the school for facilitating the experiments, and the children's parents for giving their consent.

References

1. Bellanca, J., Brandt, R.: 21st Century Skills: Rethinking How Students Learn. Solution Tree Press, Leading Edge (2010)
2. Benford, S., et al.: Designing storytelling technologies to encouraging collaboration between young children. In: Proceedings of the SIGCHI Conference on Human Factors in Computing Systems, pp. 556–563 (2000)
3. Biswas, G., Leelawong, K., Schwartz, D., Vye, N.: Learning by teaching: a new agent paradigm for educational software. Appl. Artif. Intell. **19**, 363–392 (2005)
4. Bossert, S.T.: Cooperative activities in the classroom. Rev. Res. Educ. **15**, 225–250 (1988)
5. Care, E., Griffin, P.: An approach to assessment of collaborative problem solving. Special issue: assessment in computer supported collaborative learning. Res. Pract. Technol. Enhanced Learn. **9**, 367–388 (2014)
6. Catala, A., Theune, M., Gijlers, H., Heylen, D.: Storytelling as a creative activity in the classroom. In: C&C 2017: Proceedings of the 2017 ACM SIGCHI Conference on Creativity and Cognition 2017, pp. 237–242 (2017)
7. Catala, A., Theune, M., Heylen, D.: My tablet is moving around, can I touch it? In: Proceedings of the Eleventh International Conference on Tangible, Embedded, and Embodied Interaction, pp. 495–502. TEI 2017 (2017)
8. Chandra, S., Dillenbourg, P., Paiva, A.: Classification of children's handwriting errors for the design of an educational co-writer robotic peer. In: Proceedings of the 2017 Conference on Interaction Design and Children (IDC), pp. 215–225 (2017)
9. Dillenbourg, P.: What do you mean by collaborative learning? Collaborative-learning: Cognitive and Computational Approaches, pp. 1–19 (1999)
10. Garzotto, F.: Interactive storytelling for children: a survey. Int. J. Arts Technol. (IJART) **7**(1), 5–16 (2014)

11. Hesse, F., Care, E., Buder, J., Sassenberg, K., Griffin, P.: A framework for teachable collaborative problem solving skills. In: Griffin, P., Care, E. (eds.) Assessment and Teaching of 21st Century Skills. EAIA, pp. 37–56. Springer, Dordrecht (2015). https://doi.org/10.1007/978-94-017-9395-7_2

12. Huskens, B., Palmen, A., Van der Werff, M., Lourens, T., Barakova, E.: Improving collaborative play between children with autism spectrum disorders and their siblings: the effectiveness of a robot-mediated intervention based on lego therapy. J. Autism Dev. Disord. **45**, 3746–3755 (2014). https://doi.org/10.1007/s10803-014-2326-0

13. Jamet, F., Masson, O., Jacquet, B., Stilgenbauer, J.L., Baratgin, J.: Learning by teaching with humanoid robot: a new powerful experimental tool to improve children's learning ability. J. Robot. **2018**, 11 (2018). Article ID 4578762

14. Lai, E.R.: Collaboration: A literature review. Pearson Research Report, vol. 2 (2011)

15. Leversund, A.H., Krzywinski, A., Chen, W.: Children's collaborative storytelling on a tangible multitouch tabletop. In: Streitz, N., Markopoulos, P. (eds.) DAPI 2014. LNCS, vol. 8530, pp. 142–153. Springer, Cham (2014). https://doi.org/10.1007/978-3-319-07788-8_14

16. Markopoulos, P., Bekker, M.: Interaction design and children. Interacting Comput. **15**(2), 141–149 (2003)

17. Rick, J., Marshall, P., Yuill, N.: Beyond one-size-fits-all: how interactive tabletops support collaborative learning. In: Proceedings of the 10th International Conference on Interaction Design and Children (IDC), pp. 109–117 (2011)

18. Rode, J.A., Stringer, M., Toye, E.F., Simpson, A.R., Blackwell, A.F.: Curriculum focused design. In: Proceedings of Interaction Design and Children, IDC 2003, pp. 119–126 (2003)

19. Rogat, T.K., Linnenbrink-Garcia, L.: Socially shared regulation in collaborative groups: an analysis of the interplay between quality of social regulation and group processes. Cogn. Instr. **29**(4), 375–415 (2011)

20. Roschelle, J., Teasley, S.: The construction of shared knowledge in collaborative problem solving. Computer Supported Collaborative Learning (1995)

21. Slavin, R.E.: Cooperative Learning: Theory, Research, and Practice. Prentice Hall, Hoboken (1990)

22. Thomaz, A.L., Breazeal, C.: Teachable robots: understanding human teaching behavior to build more effective robot learners. Artif. Intell. **172**, 716–737 (2008)

23. Tokic, M., Palm, G.: Value-difference based exploration: adaptive control between epsilon-greedy and softmax. In: Bach, J., Edelkamp, S. (eds.) KI 2011. LNCS (LNAI), vol. 7006, pp. 335–346. Springer, Heidelberg (2011). https://doi.org/10.1007/978-3-642-24455-1_33

24. Verhoeven, G., Catala, A., Theune, M.: Designing a playful robot application for second language learning. In: Brooks, A.L., Brooks, E., Sylla, C. (eds.) ArtsIT/DLI-2018. LNICST, vol. 265, pp. 385–394. Springer, Cham (2019). https://doi.org/10.1007/978-3-030-06134-0_42

25. Wood, D., O'Malley, C.: Collaborative learning between peers. Educ. Psychol. Pract. **11**(4), 4–9 (1996)

26. Wood, D., Wood, H., Ainsworth, S., O'Malley, C.: On becoming a tutor: toward an ontogenetic model. Cogn. Instr. **13**(4), 565–581 (1995)

27. Woodward, J., Esmaeili, S., Jain, A., Bell, J., Ruiz, J., Anthony, L.: Investigating separation of territories and activity roles in children's collaboration around table-tops. In: Proceedings ACM Human-Computer Interaction (CSCW), vol. 2 (2018)
28. Zancanaro, M., et al.: Children in the museum: an environment for collaborative storytelling. In: Stock, O., Zancanaro, M. (eds.) COGTECH. Springer, Berlin, Heidelberg (2007). https://doi.org/10.1007/3-540-68755-6_8

Designs for Innovative Learning

Learning Strategies Among Students During a Sudden Transition to Online Teaching in a PBL University

Anette Lykke Hindhede[1(✉)], Vibeke Harms Andersen[1], and Dorina Gnaur[2]

[1] Aalborg University, Copenhagen, Denmark
{anlh,vian}@hum.aau.dk
[2] Aalborg University, Aalborg, Denmark
dgn@hum.aau.dk

> *'When everything is virtual, I feel I am weightless in the universe, just like an astronaut, connected to a space station, that floats round the Earth without any grounding.'* [28]

Abstract. Increasing use of digital tools in university teaching has drawn scholarly attention to the interaction between pedagogical design and digital technologies. The accelerated transition to online learning following the COVID-19 crisis has raised several questions regarding the links between technological affordances and learning strategies, especially with regard to the role of dialogue in learning. Based on a survey of 51 postgraduate students in a Danish university with Problem Based Learning as explicit teaching strategy, where collaborative interaction and dialogue are regarded as integral to learning, this study investigates how students navigated the altered learning environment. We found that students' experiences with online teaching demonstrate reduced affordances for learning. They experienced decreased co-involvement in decision-making, decreased collaboration and a changed pedagogical setup that did not support learning through discursive meaning negotiations. Thus, whilst dialogues can be transformed by digital technology, these changes are not necessarily productive within an environment which emphasises democratic discourse. Arguably, the digital transformation will continue to evolve and influence the quality of university teaching. Our paper concludes by discussing the potential of democratic dialogic teaching to stimulate learning ecologies in online and hybrid learning environments.

Keywords: Online learning · Dialogic pedagogy · Hybrid learning · Learning ecology · Collaboration · Problem-based learning

1 Introduction and Theoretical Underpinnings

In contemporary educational thought, collaborative learning as a means to ensure collective and individual intellectual gain continues to be considered beneficial, although it

© ICST Institute for Computer Sciences, Social Informatics and Telecommunications Engineering 2021
Published by Springer Nature Switzerland AG 2021. All Rights Reserved
E. I. Brooks et al. (Eds.): DLI 2020, LNICST 366, pp. 171–186, 2021.
https://doi.org/10.1007/978-3-030-78448-5_12

may not always lead to the intended outcomes [1, 2]. Collaborative learning is defined as two or more students working together towards a shared learning goal [1, p. 247]. This is distinct from cooperation in which students work separately with portions of the task that are later fused into a comprehensive product [3]. Common to collective and collaborative learning activities is that participants are assigned a problem and asked to work together to achieve a solution [1, p. 248]. Problem-based learning (PBL) is a pedagogical design in which students in groups engage actively in problem-solving and thereby extend their knowledge and co-create a solution to the given problem. In this collaborative activity, inquiry entails the critical analysis and synthesis of new information gained when confronting opposing discourses. This meta-reflection as a collaborative practice in which students develop dialogic argumentative thinking and writing competences reflects the core of dialogic pedagogy [4, 5], which values the heterogeneity of knowledge, of language and of reasoning [6]. An *epistemological* definition of dialogue is that it offers a theory of meaning. It implies that education should be designed to engage students in an ongoing process of shared inquiry that takes the form of a dialogue [7]. An *ontological* definition of dialogue is concerned with the very nature of our existence. Dialogue is not only a means to construct knowledge between selves, rather, selves and reality are part of the dialogue. These two different understandings of dialogic education focus on transforming the self, reality or social reality [8].

Dialogic pedagogy has primarily been described in relation to school contexts and with a focus on classroom teaching. Here, it is deemed as a participatory imperative [9, p. 1998] as it aims to develop and legitimise participation from all parties in classroom interactions [10]. Only a few studies have described dialogic pedagogy in online contexts. One of these is Simpson [11, p. 136], who argues that when social networks are added to the learning context, then it is the teacher's responsibility, 'through the explicit and focused use of critical reflective dialogic practices' in activities aligned with required outcomes, to provoke students to value the role of dialogue in their own learning. Thus, the teacher is active and responsible for enhancing learning. However, democratic pedagogical purposes can be challenging. Aiming to foster each student's agency and participation in the construction of shared classroom understandings means that all students should be positioned as capable collaborators. The idea that humans through education develop their capabilities of agentic and autonomous action and that education is key to empowerment and emancipation has been central since the Enlightenment [12, 13]. Whereas agency in past decades has been theorised rather one-sidedly in the structure-agency debate, as have the factors that promote or hinder agency [14–16], in this paper we focus on the ecological conditions which impact the achievement of agency. This action—theory approach is based on Dewey's so-called transactional constructivism [17, 18] meaning that the transaction between organism and environment can be read as an account of the construction processes that lie beneath all human activity. Within this understanding, agency is seen as the way in which actors 'critically shape their responses to problematic situations' [12, p. 11]. Thus, rather than understanding agency as something that resides in the individual, in this paper we see agency as something emerging from the interaction between humans (here students and teachers) and situations (here the educational context).

Biesta [19] focuses on what might be involved in education in a world of human difference. He argues that 'good' education requires answering the question of purpose which, according to him, relates to the three domains of qualification, socialization and subjectification. Qualification consists of knowledge, skills and dispositions. Socialisation is about the formal and informal ways in which we, through education, become part of certain social, cultural and political orders. In his description of the concept of subjectification, Biesta takes as his point of departure Kant's concept of formation, i.e. the way in which through upbringing and teaching the individual can become able to make use of their mind without the guidance of another and thus step out of their self-inflicted immaturity. It requires teachers to maintain a sense of their students as unique individuals in the process of developing their individual voices and relationships with the world. However, while research in classroom dialogue is well established [9, p. 1996], recent technological advances have led to a scholarly interest in the mediating role of digital technologies in enabling collective knowledge building in education [34]. Ordinary conceptions of dialogue can be extended to include digital technologies in a dialogic-interactive manner [20]. In the age of the Internet, the variety of learning trajectories has increased significantly, spanning physical and virtual domains and involving an increasingly personalised use of digital technologies.

An interesting concept in this connection is that of learning ecologies [21–23]. Learning ecologies are instances of eco-social systems that follow a developmental trajectory, i.e. they involve relational and interactional processes in the construction, sharing, and reconstruction of meaning. Seen from a learning ecology perspective, the Internet and digital technologies have altered our environment and hence the way we learn. Related to this is the concept of hybrid learning, which involves dialogue across a variety of learning and practice settings through different types of physical and digital networks and various technologies. Hybrid learning supports 'fluid forms of becoming and being in, with and for the world' [24, p. 78]. Digital technologies play a crucial role in connecting and mediating learning expeditions in hybrid environments as they afford various ways of engaging with technology, which serve individual as well as joint objectives, i.e. interaction with digitised content, problem-solving, dialogue and relationship building and discursive practices involving various forms of dialogue [25, 26].

Like any ecological system, a learning ecology can function in a balanced and sustainable way, but it can also destabilise due to uninformed action or disruptive events. The forced transition to online teaching due to the COVID-19 lockdown has caused such a disturbance. The sudden transition to online teaching has forced educators and students to skip the expected gradual transformation towards increased digitisation in university teaching, making all teaching activities 100% online overnight. In this paper, we are interested in how this transition has affected the dialogic element in teaching and learning, and what the role of dialogue would be in recovering the ecological learning balance in an online and possibly hybrid learning environment. We focus on students' experiences in order to get a sense of the individual and collective discourses that inform students' perceptions and judgements on what motivates and drives their action during the unexpected digital transition due to the COVID-19 crisis, particularly regarding the role of technology in supporting or inhibiting discursive spaces through dialogue. We find this context very fruitful for considering the agentic dimension of students' actions as their responses reflect changing orientations due to new structural environments of action due to the COVID-19 pandemic.

In order to gain insight into the critical aspects, challenges and potentials of collaborative learning and dialogic pedagogy with the rapid transition to digital teaching, we ask:

> How is the accelerated digital reorganisation of teaching experienced by a group of students in a PBL university in terms of collaboration, dialogue, and experienced learning? What educational potential can be found in digital learning ecologies for enhancing the three domains of the purpose of education (qualification, socialisation and subjectification)?

2 Methods

The present study is based on data from a master's degree programme in the field of humanities in a PBL university during the first three months of the COVID-19 lockdown. It aims to map students' learning experiences during the digital turnover. The digitised experience covered online course teaching such as synchronous and asynchronous lectures. Teachers were free to choose from among the platforms that the university had cleared under the General Data Protection Regulation (GDPR), which included Zoom, Teams and Skype for Business. Some teachers also chose to communicate only via Moodle, the Content Management System used by the university, where they uploaded PowerPoint presentations with voiceovers. The online supervision of student PBL projects was conducted in one of the aforementioned GDPR-compliant platforms, i.e. Zoom, Teams, or Skype for Business. Most students did collaborative group work, although some chose to work on their own. Teachers were free to design their teaching in any way they preferred, so the students were exposed to different forms of online teaching. This master's degree programme enrols students from diverse backgrounds. Some have academic bachelor's degrees in humanities or social sciences fields and are relatively well-versed in working and writing academically. Others have professional bachelor's degrees, which often focus more on training for a particular profession and its related methodologies. A postgraduate degree is a challenging experience for many students, especially during the main PBL project each semester, in which they have to collaborate across various academic traditions. A survey was developed by the authors which addressed a variety of digitised course teaching activities that the students were exposed to during the lockdown, combined with prior experiences from the time of their bachelor's studies; the survey also related to the project work with regard to online supervision and online group collaboration. We hypothesise that students' perceptions of these issues reflect their level of agency as a configuration of influences from the past (experiences with teaching from before the lockdown), orientations towards the future (what enables learning), and their engagement in the present [3]. The online survey was answered by 51 out of 136 active students (i.e. 37.1%); responses represented the university's two campuses relatively equally.

We supplement the survey results with two student testimonies retrieved from student-conducted projects that also focused on the sudden exposure to online teaching [27, 28]. Both student projects were based on a combination of surveys and qualitative interviews. One study was based on the same master's degree programme as the present study [27], but only at one campus. The other was aimed at the broader student body at the PBL university [28]. In addition, we draw on a written student reflection during the lockdown, submitted to one of the researchers, who was assigned as the student's learning supervisor [29].

The data has been analysed drawing on Biesta's three separate educational dimensions of *qualification, socialisation*, and *subjectification*. The questions in the quantitative part of the survey (see Tables 1 and 2) provide data that reflect the concept of *qualification*, addressing if and how online education provided students with subject knowledge (i.e., ultimately helped them become critical thinkers and thus qualified their participation in societal affairs). The qualitative part gathered students' open-ended responses to what supported learning, whereby *socialisation* – which is typically formulated in the overall curriculum – is reflected in students' responses to the values and behaviours that the university, according to them, is meant to foster. Finally, *subjectification* is investigated by focusing on statements that reflect the teachers' ability to sustain a sense of the students as unique individuals by helping them to develop their individual voices and relationships with the world. We are particularly interested in the possibilities for synergy between the three domains, but also where they are in conflict. This, we argue, will enable us to deal with 'trade-offs' between the three domains [14, p. 79].

3 Results

The quantitative part of the survey included questions regarding the degree of ease in transitioning to an online teaching environment as well as online PBL supervision, primarily relating to the domain of *qualification* (the ability to acquire new knowledge and skills in the new online learning environment). Responses were expressed on a Likert scale with five possible degrees, where (1) is 'Totally disagree' and (5) is 'Totally agree'. Table 1 presents a summary of the questions and the distribution of the answers within three categories, where (1) and (2) combine into 'Mostly disagree'; and (4) and (5) comprise 'Mostly agree'.

Table 1. Questions related to the digitised teaching and online PBL supervision

Question	Mostly disagree	Neutral	Mostly agree
It was easy for me to get used to the online teaching	33%	25%	42%
It was easy for me to keep up with the communication about the organisation of teaching, including presentation of content, tasks and activities, etc	40%	36%	24%
It was easy for me to get acquainted with the various digital formats and software used to conduct teaching online	31%	36%	33%
It was easy for me to adapt the provision of online teaching to my own learning preferences	42%	29%	29%

(*continued*)

Table 1. (*continued*)

Question	Mostly disagree	Neutral	Mostly agree
I received help and answers to my inquiries from the teachers regarding the online teaching	7%	29%	64%
It was easy for me to get used to supervision going on online	11%	4%	85%
It was easy for me to select suitable technologies for online supervision	7%	15%	78%

There is a relatively wide distribution of perceived degree of ease regarding various aspects of the online teaching. Elements relating to communication about the logistics of online teaching as well as to the ease of adapting digitised teaching options to individual learning preferences seem critical to many of the students, possibly reflecting uneven opportunities for learning. This group of students are accustomed to classroom teaching and it may be confusing to find one's place in a multimodal environment with different affordances for learning. Their responses reflect difficulties in navigating this environment and the need for teacher control. Notably, teachers are appreciated for offering clarification in response to students' uncertainties, which indicates that from the students' perspective, teachers play a strong role in guiding students to become part of this new social order. This might indicate that during the lockdown, most students expressed agency in asking their teachers for help. Remarkably, students seemed to adapt easily to an online PBL supervision environment and were able to select suitable technologies. This could partly reflect a prior *socialisation* to the more autonomous nature of PBL project work, and partly the more manageable coordination task regarding online interaction around a known subject, i.e. students' projects, with only one teacher/supervisor. However, in terms of collaboration, we see that students prefer getting feedback from the teacher rather than peers.

In the survey, we also asked students about which ways of digitising teaching they considered important to their learning. In responding to this question, they had the option to select from several pre-formulated options as well as add their own statements. The responses are summarised in Table 2, which lists the distribution of responses starting with the issue deemed most important. Overall, we see that students ask for explicit criteria for their work during online teaching, addressing the dimension of education that focuses on *qualification*. Here, we see that students deem it very important to get explicit norms from teachers as to what is considered the 'right' student behaviour. This raises the question of whether presenting students with predefined standards for the 'right' knowledge (cf. the *qualification* domain) and the 'right' behaviour (cf. the *socialisation* domain), followed up by narrowly structured and objective-oriented education, is conducive to learning. Alternatively, exposing students to less-structured contexts for knowledge creation, which allow a more fluid process of knowing similar to moral or political education, could elicit the students' active engagement and sense of learning as agentic action (cf. the *subjectification* domain). Qualifying students with planned knowledge represents but a limited educational aim. From this perspective, it is alarming that only 24% of students expected to be part of decision-making processes, whereas 42% preferred real-time activities during the online teaching which requires standard sequences of activities unless the teacher is able to improvise in response to each class' unique flow (for further discussion on the artful balance of structure and improvisation, see Sawyer (30)).

Table 2. Overview of students' answers on the approaches to digitising teaching that are important to their learning

Following ways of digitising teaching are important to my learning	
The expectations of students in online teaching are clearly communicated	76%
Adding audio to a PowerPoint presentation to display as needed	64%
I receive online feedback from the teacher on my assignments	60%
The online teaching is organised so that I develop my critical, analytical, creative or practical sense of working with the material	53%
I can structure my time as a student in an online teaching environment	51%
The teacher makes a plan for online teaching, which helps me structure my time	47%
Online teaching is organised so that I experience increased self-insight and can reflect on and assess my own experiences and knowledge	44%
I get activated to participate individually in real-time teaching	44%
I get activated to work in groups in real-time online teaching	42%
There is a uniform way of organising online teaching across the various modules	42%
The lectures are synchronous, using online meetings with direct presentation	35%
I participate in group assignments to solve in a set group before or after teaching	31%
I experience being seen and recognised in an online teaching environment	29%
The online teaching supports that we know & contribute to each other's learning	24%
The online teaching is organised so that I feel part of the decision-making processes	24%
I receive online feedback from my fellow students	22%

In the qualitative part of the questionnaire, we see a more mixed picture. Here, despite the new structural circumstances, in terms of the domain of *socialisation,* students express their expectations that teaching be based on principles of a high degree of student involvement. This is exemplified by the following statement made in response to being asked about what enhances learning:

'That we as students are involved as much as possible, despite the online format. For example, by organising and facilitating parts of the teaching itself so that you get co-responsibility, feel ownership and commitment in the digital space, where you can otherwise easily become invisible'.

This student is constructing a distinction between analogue and digital teaching and arguing that, in the case of the latter, when aiming for the achievement of digital *qualifications,* there is a risk of a negative impact on the domain of *subjectification.* There is a perceived risk of becoming invisible as students – not only to the teacher/supervisor, but also to other students, and to themselves [33]. In digital teaching, teachers often ask students to join meetings without video and audio enabled unless asked to say something. Thus students can maintain a visible digital presence through their logged-in status but are physically and mentally non-present. Student involvement depends even more on the teacher to be aware and to let the students engage actively with the learning situation. The analogue learning environment is endowed with known modalities to escape feeling invisible –students can choose to involve themselves and teachers and the other students can enable and elicit involvement. The students mention how they used to learn from listening to other students' discussions, and by asking questions. This is made more difficult in the digital learning environment.

Another student emphasises how group work related to teaching sessions is considered of worth for his/her learning: 'It helps me to be part of a group, I have greatly benefited from my project group'. To this student, there seems to be synergy in the domain of *qualification* and *socialisation* when working in groups as not only qualification and academic achievements count for her. This is also a goal in the pedagogical design of PBL, and although group work was made digital too, belonging to a group as a place for qualification, socialisation and subjectification may still be important.

A qualitative investigation conducted by some of the participating students [27] demonstrated other viewpoints. Here, some students explained how they chose to work with peers and groups other than the groups generated by the teacher. They chose fellow students who they knew, who were visible, and who were considered as being more actively engaged in the teaching sessions and showed agency in learning. They also emphasised the importance of the teacher/supervisor's participation in online discussions.

However, according to many of the students, group work ought to be orchestrated in a certain way to be considered successful. This student explained how she expected the teacher to take the role of facilitator during their group work:

'It helped if you met in small groups, that is, if the groups were formed by the teacher. In those cases, it was instructive that the teacher went in and out of the small digital forums and participated in discussions'.

According to Biesta, teacher judgment is essential to education. With the teacher as a facilitator, key educational questions of content, purpose and relationships might be forgotten. Thus, to Biesta the language of learning is insufficient for expressing what education is about because it does not answer the questions of learning 'of what' and 'for what' [14, p. 76]. However, we do not know from the student's quote the way in which the particular teacher participated in the discussions. It might be that they actually acted in a way which promoted agency for both the teacher and the students.

Students' responses also describe experiences with using diverse platforms for group work and dialogues, reflecting how students use dialogue to scaffold their learning and the importance of the ability of the digital tools to create a shared dialogic space. This is reflected in the following quote, in which a student explains what increases her learning:

'That it is possible to do reflection tasks and other tasks with fellow students, so that the material is processed in different ways than reading on your own and attending lectures. The timeframe in which the work is to take place should be fixed'.

Students appeared to struggle to maintain a stable identity when the only contact with peers was through digital technologies. Thus, they seemed to orient themselves toward a change to the existing order so that different ways of being became possible. An example of this attempt to be a subject of action is reflected in the following quote, where a student explains what is key to online teaching:

'The most important thing is the contact with other people. So, there must be good opportunities to discuss and talk to one another; both teachers and fellow students, so that one does not sit alone with one's thoughts'.

There are inescapable differences between the opportunities in a more conventional classroom and those online. It seems more difficult to reach a level of depth of reflection and critical thinking during online teaching. The statement below represents dominant issues in students' perceptions of missing components of the experience, the lack of which inhibits their learning:

'I have experienced missing the dialogues that typically occur in the classroom – between teacher and student – which to me greatly help to comprehend the material that is read in advance and which is elaborated during lectures by the teacher. It is as if a "layer" is missing – which we have not been able to access during this period'.

What we see here is an example of a missed opportunity to interact with more knowledgeable others as compared to the previous social order, i.e. live lectures. On the other hand, merely transposing familiar structures to a digital format is not an option as it inhibits the construction of meaning through direct dialogue. Rather, new and explicit active learning strategies are required in order to foster learning, which is explained by a student as follows:

'The online format makes it necessary to create even more variety. Although traditional lecturing for a long period is also tiring in real-time teaching, it is even more tiring in virtual spaces. Therefore, the passive listening in front of the screen really impedes my learning. Things need to be staged and exercises implemented to a much higher degree when it is virtual'.

Several quotes reflect the student demand for more variety in order for digitised teaching to become more engaging. Although this aspect is considered central in ordinary class teaching, in online teaching it becomes even more crucial. The data, however, underscores that technological variation in itself does not ensure the quality of learning if the students have the feeling of being left to themselves, as the title of an abovementioned study puts it: Am I alone here – or are there others with me?' [27]. More specifically, students ask for dialogue practices that are designed to spur curiosity in learning and promote critical reflective thinking.

In our survey, we also asked students about what enables and impedes learning when working together online in project groups. Generally, the students deplore the lack of face-to-face contact in project groups, which severs the exercise of individual agency due to difficulties of mutual relating. As one student explains:

'The physical isolation. It is difficult to work in project groups without face-to-face contact. It's not the same. Physical presence cannot be replaced by digital collaboration'.

Another student emphasises how true dialogue is missing during online interaction:

'We missed the dialogue. It's just not the same talking over an online media compared to talking face to face'.

In the pedagogical design of PBL, dialogue is a central component of collaborative learning, enabling the students to engage in common problem-solving. They challenge, provoke and argue as a means of knowledge negotiation in order to find new ways of understanding, which integrate their different academic and practical experiences. These types of dialogue are difficult for students to maintain online, especially when they do not know each other beforehand.

The statements below represent key issues of what enables learning, and interestingly, students relate to both domains of *qualification* and *socialisation* in their responses:

'Virtual meetings with the opportunity to *see* other group members and supervisors are central'. Moreover, 'strict appointments, committed group members, more frequent supervision enable learning'. Another student stated: 'Written feedback prior to the online meeting. Well-prepared supervision'.

However, there are many factors related to the online digital environment which are experienced as restricting or challenging learning in project groups as students' ability to interact is challenged:

'I think it hampers my learning in the project that we are not sitting together. For me, it is especially about not sitting in the same room and having the interaction you would if sitting together physically'.

Students also face new technical challenges which arise as a result of digital learning. These relate to their having to sit in front of a screen for many hours, that the internet connection might fail, and, maybe the most important, that they have difficulties reading each other's body language. Therefore, it became easy to misunderstand signals:

'Technical challenges may occur, and meetings are less dynamic because it is more difficult to read each other's cues and signals. More specifically, lack of body language. It is weird and not always optimal to sit and communicate on Skype'.

'It may be harder to facilitate good discussions/workshops around topics – it is also harder to sit in front of a screen instead of meeting physically. It is harder to create relationships in a group when you only meet digitally'.

On the other hand, students seem to be able to see beyond the limiting factors of technology and refer to how online teaching and project work can improve their respect for each other and be more effective. Virtual collaboration related to project work seems to increase the necessity of better planning as well as mutual alignment of expectations and a certain discipline in presenting arguments and sticking to the point, especially as body language is no longer a possible way to argue:

'It creates some flexibility, but it requires the group to understand that this is how we work. In our group, we established the expectation that we work when it fits in each other's calendar'.

'We are more efficient, and aware of respecting each other's time. We have also become very good at handing out tasks, and then continually use the [online] platform to discuss things briefly. You are also forced to become clearer in how to articulate because you do not have gesticulation, a whiteboard or other artifacts to use in your explanation and communication with fellow students'.

However, many obstacles turn up within group work due to the total transition to online communication. As the group of students is very diverse, not all have strong competences in argumentation as an acquired skill helping them to move from pure disagreement to evidentiary discussion. As Alexander [31] argues, it is hardly surprising that some students view argument as conflict, and this becomes an even more sensitive problem to tackle in an online format. Students seem to employ evasive strategies to avoid professional disagreements that might escalate to personal conflicts, as the virtual environment is perceived to pose obvious limitations to conflict resolution. The challenge to reaching intellectually stimulating relationships in the virtual space was described in various ways:

'That not everyone contributes equally, which creates an overload for oneself and means that one does not have anyone to discuss with anyway in the meetings. Therefore, you lose the sense of how much people read and prepare at home. In tactical terms, I also bend off if something could develop to a small conflict or discussion that would otherwise be exciting and instructive to unfold at physical meetings. Nevertheless, I have chosen this strategy to stick to what is possible when communicating through the computer, and to avoid misunderstandings. This is also about the lack of body language to help navigate, despite the camera'.

'That you do not know your group members beforehand and cannot build relationships in the same way as if you could physically "hang out". It is, therefore, 'difficult to create relationships'.

'It can be difficult to communicate over an online platform when you don't know each other. Because you do not have the same group dynamics as when you sit together physically – you do not really get to know each other, and misunderstandings can easily arise.'

According to Biesta, teacher involvement and judgment are essential to education. We see in our survey that students are very explicit about what they find important to their learning regarding teachers/supervisors. They want clear information about the expectations of students and the responsibility for organising teaching so that they can develop a critical, analytical, creative or practical sense of working with their tasks. In addition, they ask for plans that help them structure their time, to be engaged fully in the learning process, and that their supervisor plays the role of facilitator during group work and gives well-prepared written feedback prior to the online meetings.

During the COVID-19 lockdown, many teachers spent much more time than usual on teaching and supervision in order to compensate for the loss of physical meetings and because they had to change their normal classes to an online format overnight. Transitioning to online teaching opens up important discussions concerning the position of the teacher/supervisor and the time spent on readjusting the learning design when traversing the analogue and online teaching spaces. Rather than covering both modes of teaching, teacher judgement might well be applied to combining the two modes into a hybrid learning space wherein they become gradually more integrated.

Biesta, building on the work of Dewey, states that processes of subjectification require that the teacher aims to support all students to become intellectually independent and responsible 'subjects of action' [32, p. 64]. The above testimonies may be interpreted as troubling episodes with misunderstandings that can be transformed into learning opportunities for all.

4 Discussion

In their review of the factors that influence learning outcomes, student satisfaction and collaborative engagement in e-learning and blended learning, Nortvig et al. [33] found that the most prominent influences were the teacher's presence in online settings; interactions between students, teachers and content; and designed connections between online and offline activities as well as between campus-related and practice-related activities. In this paper, we found that students expressed a desire to be exposed to education that was safe and predictable. This is not surprising, taking into consideration the context of the educational changes. Regarding the potential for collaborative activities, dialogic exchanges and discursive meaning creation, the students experienced severe limitations in the online context as compared to the previous socio-pedagogic framework. They deplored the lack of face-to-face interaction, yet did not seem to exert their agentic capacity for co-constructing meaning in an online environment, presumably due to experiencing increased complexity. However, this leaves us with the question of how to plan digitally supported online teaching while at the same time maintain an orientation towards the independent thought and autonomy of the ones being educated. According to Biesta, teachers should also enact agency, that is, the freedom to act independently of the determining constraints of social structure [14]. If we accept the premise that education always impacts on the three domains of qualification, socialisation and subjectification, then, following Biesta [19, p. 77], 'as educators we must take responsibility for what it is we seek to achieve in each of these domains'. During the COVID-19 crisis, teachers were not able to fully exercise their judgment about the appropriateness of how they teach and organise their educational efforts. Rather, this was dictated by the government.

What, then, are the implications for students' agency? Biesta asks 'Is it indeed a good idea to treat students as customers and give them what they want? Does this give them a much needed 'voice' in the educational process and does it therefore enhance the overall quality of the educational endeavour?' [14 p. 82]. It is crucial to understand that students and teachers have very different voices because of their different responsibilities and expectations. So, how do we construct a dialogical pedagogy in a way that includes different voices?

Regarding the application of ecological learning principles in higher education, Richardson [23] has pointed at three main priorities. First, an ecology of learning should offer rich possibilities to locate and access learning items that students can organise and interact with to satisfy learning needs. Second, a learning ecology must support social learning through collaborative learning activities, wherein students engage in group discussions to explore content and 'discuss and share insights within their specialised communities of practice' [23, p. 48]. Third, a learning ecology requires an intentional design to create 'a learning system that adapts to varying student needs' [23, p. 48]. The perspective of learning ecologies suggests the need to offer rich and diverse possibilities for students to access learning in accordance with their needs and preferences in order to attain *qualification.* Equally important is *socialisation,* for which collaboration and discursive meaning creation through dialogue are paramount. Finally, learning ecologies require intentional designs for learning that take advantage of the rich media ecology and the hybridity of learning spaces to inform student action and promote mutual engagement. A learning ecology can thus accommodate the *subjectification* function, spurring students towards the autonomy of thought and action. The dialogic element seems to be crucial in order to facilitate interaction and successful communication as well as invite students to value a diversity of perspectives and various ways of knowing. A learning ecology design should include explicit opportunities to engage in critical reflective dialogic practices in alignment with the learning aims.

In a PBL environment, dialogic practices can take on a variety of forms, depending on the learning activity, whether related to teaching, and thus to declarative and procedural knowledge, or to independent project work, which capitalises on the construction of conditional and functional knowledge. Similarly, dialogic activities ought to circulate reflective and meaning-creation processes across various learning environments.

Such activities could involve students or groups of students and the teacher/project supervisor, but also draw on input from external collaborators from the various organisational contexts where student projects are hosted as well as relevant formal and informal student and teacher networks. A digitally enhanced media ecology accommodates multimodal practices and thus rich possibilities to frame knowledge exchanges, e.g. written, pictorial, audio and video-based. More significantly, it also expands and enhances new discursive practices in hybrid, cross-learning and action spaces, connecting people and meaning in various situations, both synchronously and asynchronously. Dialogue needs to be specifically framed to suit such an expanded hybrid learning environment, which transcends familiar classroom practices. This can be a challenge not only for students, which explains the restrictive use of online engagement, but also for teachers, who need to redirect their pedagogic thinking from content-centric to a dialogical and discursive orientation of knowledge and practice.

Whether experiences with the sudden transition to online teaching due to the COVID-19 crisis can provide a reliable point of departure for designing for learning in a hybrid learning environment is a point for further discussion. In fact, as argued elsewhere [16], in this paper we, too, have found that students' responses reflect the temporariness of contexts of action and thus that their agentic orientations are subject to substantial variation. Their responses, therefore, display how they move between different and unfolding contexts that make them orient or 'recompose' the situation in various ways. The new emerging online learning environment is sustained by, and at the same time potentially altered through, students' agency as a response to teaching strategies. Similarly, teachers' actions are subject to students' learning affordances in the new environment. Dialogue and eliciting student co-involvement at a meta level, regarding the potentiality of digitally extended learning ecologies, may prove central to designing for learning in new hybrid spaces.

This study is exemplary, since many other higher education institutions may have had similar experiences with transitioning to online teaching during the COVID-19 lockdown. Therefore, we believe that knowledge generated through this study regarding the role of dialogic practices – not only to enhance learning and promote critical reflective thinking but also in co-designing learning in online hybrid spaces – can inform others in the field. We may not base our future strategies for digitisation in higher education on the experiences from this sudden online transition, but we may use these experiences to initiate an exploration of dialogic practices to design for digitally enhanced learning ecologies. It would be of great interest to compare our results with those seen at other higher education institutions. However, at the time of writing, the basis for such a comparison does not yet exist as many of the studies from the period of the lockdown are not yet available.

5 Conclusion

In this paper, we focused on how a group of students at a PBL university have navigated the altered learning environment during the COVID-19 pandemic in terms of collaboration, dialogue, and experienced learning. We drew on an ecological conception of agency-as-achievement. We found that students' agency is achieved when they are able to ask for help from their teachers. In terms of collaboration, we also saw that students prefer getting feedback from teachers rather than peers. Students ask for explicit criteria for their work during online teaching, emphasising the dimension of education that focuses on qualification. Here, we see that students deem it very important to get explicit norms from the teachers as to what is considered the 'right' student behaviour.

Furthermore, we investigated what educational potential can be found regarding the three domains of the purpose of education identified by Biesta (qualification, socialisation and subjectification) in digitally enhanced learning ecologies. While students prefer teacher control in terms of predefined 'right' knowledge and social behaviour, informed judgement might encourage teachers to expose students to more open forms of knowing, and thus the challenge of withholding uncertainty. Students' qualitative responses cover a more mixed picture: Despite the new structural circumstances, in terms of the domain of *socialisation,* students expressed their expectation that the teaching is based on principles of a high degree of student involvement. When aiming for the achievement of *qualification* in a digitised online environment, there is a risk of a negative impact on the domain of *subjectification* as the students might become invisible. To some students, there seems to be synergy in the domain of *qualification* and *socialisation* when working in project groups, where not only qualification and academic achievements count.

Students' agency is hindered in a setting where their only contact with peers is through digital technology. In and through the particular ecological condition and circumstance of complete digitisation, it seems more difficult to reach deep reflections and critical thinking. Students' agency is also hindered if levels of variation in the digitised teaching, involving and activating are low. During online project work, the students miss face-to-face contact, which severs the exercise of individual agency in collaborative processes due to difficulties of mutual relating. The students mentioned physical isolation, loss of dialogue, loss of interaction and employing strategies to avoid professional disagreement because it can be difficult to separate these from personal conflict.

The ideal of PBL and group work is that students should be challenged to wrestle meaningfully with uncertainties that bear conflicting perspectives. Although democratic habits of mind might be achieved in practices where students learn to recognise and respect opposing arguments and differences of perspectives as a path to argumentative competence, in a COVID-19 crisis, this may be a learning goal too high to aim for.

It is important to take into account that the results obtained are shaped both by the feelings generated by the forced transition to the teaching modality and by the inexperience of teachers and students in developing activities in online contexts. This inexperience will also have certainly influenced the students' perceptions and even their performance in learning activities. The conclusions of this study must, therefore, be weighed according to the circumstances.

References

1. Jeong, H., Hmelo-Silver, C.E.: Seven affordances of computer-supported collaborative learning: how to support collaborative learning? how can technologies help? Educ. Psychol. **51**(2), 247–265 (2016)
2. Kuhn, D.: Thinking together and alone. Educ. Res. **44**(1), 46–53 (2015). https://doi.org/10.3102/0013189X15569530
3. Dillenbourg, P.: Collaborative Learning: Cognitive and Computational Approaches. Elsevier Science Inc, New York (1999)
4. Alexander, R.: Towards dialogic teaching: Rethinking classroom talk. (4th ed.). Thirsk, Dialogos (2008)
5. Mercer, N., Littleton, K.: Dialogue and the Development of Children's Thinking: a Sociocultural Approach. Routledge. (2007)
6. Mayer, S.J., O'Connor, C., Lefstein, A.: Distinctively democratic discourse in classrooms. In: Mercer, N., Wegerif, R., Major, L. (eds.). The Routledge International Handbook of Research on Dialogic Education. London: Routledge (2019)
7. Linell, P.: Rethinking Language, Mind, and World Dialogically. IAP, Charlotte (2009)
8. Mercer, N., Wegerif, R., Major, L.: The Routledge International Handbook of Research on Dialogic Education. London: Routledge (2019)
9. Major, L., Warwick, P., Rasmussen, I., Ludvigsen, S., Cook, V.: Classroom dialogue and digital technologies: a scoping review. Educ. Inf. Technol. **23**(5), 1995–2028 (2018). https://doi.org/10.1007/s10639-018-9701-y
10. Mercer, N.: The social brain, language, and goal-directed collective thinking: a social conception of cognition and its implications for understanding how we think, teach, and learn. Educ. Psychol. **48**(3), 148–168 (2013)
11. Simpson, A.: Designing pedagogic strategies for dialogic learning in higher education. Technol. Pedagog. Educ. **25**(2), 135–151 (2016)
12. Biesta, G., Tedder. M.: How is agency possible? Towards an ecological understanding of agency-as-achievement Working Paper 5. Exeter: The Learning Lives Project (2006)
13. Freire, P.: Pedagogy of the Oppressed. Penguin, London (1970)
14. Biesta, G.: What is education for? On good education, teacher judgement, and educational professionalism. Eur. J. Educ. **50**(1), 75–87 (2015)
15. Mezirow, J.: Transformative Dimensions of Adult Learning. Jossey-Bass, San Francisco (1991)
16. Emirbayer, M., Mische, A.: What is agency? Am. J. Sociol. **103**(4), 962–1023 (1998)
17. Vanderstraeten, R.: Dewey's transactional constructivism. J. Philos. Educ. **36**(2), 233–246 (2002)
18. Dewey, J.: Democracy and Education. Free Press, New York (1944/1916)
19. Biesta, G.J.: Good Education in an Age of Measurement: Ethics, Politics, Democracy. Routledge, London (2015)
20. Twiner, A., Coffin, C., Littleton, K., Whitelock, D.: Multimodality, orchestration and participation in the context of classroom use of the interactive whiteboard: a discussion. Technol. Pedagog. Educ. **19**(2), 211 (2010)

21. Jackson, N.: Exploring learning ecologies. Lulu. com. (2012)
22. Barnett, R.: The Ecological University: A Feasible Utopia. Routledge. (2017)
23. Richardson, A.: An ecology of learning and the role of e-learning in the learning environment. Global Summit Online Knowl. Netw. 47–51 (2002)
24. Hilli, C., Nørgård, R.T., Aaen, J.H.: Designing hybrid learning spaces in higher education. Dansk Universitetspædagogisk Tidsskrift **15**(27), 66–82 (2019)
25. Stommel, J.: Hybridity part 2, what is hybrid pedagogy? Hybrid Pedagogy (2012). https://hybridpedagogy.org/hybridity-pt-2-what-is-hybrid-pedagogy. Accessed 15 July 2020
26. Aaen, J.H., Nørgård, R.T.: Participatory academic communities: a transdisciplinary perspective on participation in education beyond the institution. Conjunc. Transdisciplinary J. Cult. Participation **2**(2), 1 (2015)
27. Nielsen, M., et al.: Er jeg alene her, eller er der andre med? Digital undervisning på Aalborg Universitet, København under COVID-19 (Am I alone here, or are there others with me? Digital teaching at Aalborg University during COVID-19). Unpublished project report, Learning and Change Processes, Aalborg University (2020)
28. Brohammer Hansen, S.L., Blachman, S.: En evalueringsundersøgelse over digitale erfaringer gennem undervisningen af valgfagene på 8. semester (An evaluation study of digital experiences of the teaching in elective modules of 8th semester). Unpublished project report, Learning and Change Processes, Aalborg University (2020)
29. Submitted written student learning reflection to her supervisor during COVID-19 (2020)
30. Sawyer, R.K.: What makes good teachers great? The artful balance of structure and improvisation. Structure and improvisation in creative teaching, In: Sawyer, R.K. (ed.). Structure and improvisation in creative teaching. Cambridge University Press, pp. 1–24 (2011)
31. Alexander, R.: Whose discourse? Dialogic Pedagogy for a post-truth world. Dialogic Pedagogy Int. Online J. **7** (2019)
32. Biesta, G.J.: Beautiful Risk of Education. Paradigm Publishers, Boulder (2013)
33. Nortvig, A.M., Petersen, A.K., Balle, S.H.: A literature review of the factors influencing e-learning and blended learning in relation to learning outcome, student satisfaction and engagement. Electron. J. e-Learn. **16**(1), 46–55 (2018)
34. Major, L., Warwick, P.: 'Affordances for dialogue': the role of digital technology in supporting productive classroom talk. In: The Routledge International Handbook of Research on Dialogic Education. Routledge, London (2019)

Increasing Reading Engagement for Danish Gymnasium Students: *The Hosier and His Daughter* as a Serious Game

Mads Strømberg Petersen, Niklas Lee Skjold Hansen, Gustav Jakobsen, and Thomas Bjørner$^{(\boxtimes)}$ (iD)

Department of Architecture, Design and Media Technology, Aalborg University, A.C. Meyers Vænge 15, 2450 Copenhagen SV, Denmark
tbj@create.aau.dk

Abstract. This study outlines how a serious game was implemented using transmedia storytelling to engage students in a Danish gymnasium when reading the mandatory novella *The Hosier and His Daughter*, written by the Danish author St. St. Blicher in 1829. The study is based on 52 students from two gymnasium Danish classes. The study's novelty and importance lie in its focus on using a participatory design approach to involve the teachers as co-designers at a very early stage. The transmedia setup was based on the following procedure: read seven pages of the story in the textbook, play seven pages as a game that includes reading and voice-overs and then read the rest of the story in the textbook. A formative evaluation was administered using a questionnaire after the first reading and after the gameplay. Furthermore, there were in-depth interviews with both teachers and students. The findings indicate that the serious game improved reading engagement, leading to much higher immersion levels, ease of reading and enjoyment of reading the story. The story in the game was well told, and the learning outcome was achieved through increased engagement.

Keywords: Reading engagement · Serious game · Transmedia storytelling · Co-designers

1 Introduction

This study was aimed at using serious gaming as an approach to engaging students enrolled at Ørestad Gymnasium (Copenhagen) in reading Steen Steensen Blicher's 1829 novella *The Hosier and His Daughter* (*Hosekræmmeren*) [1]. The story resembles that of *Romeo and Juliet* in that it involves young star-crossed lovers who cannot be together and ends in tragedy. The story employs poetic realism, but it is not easy to read. This is partially due its early-nineteenth-century wording and writing style, but its complexity mainly arises from the story's narration, which presents potential barriers to reading engagement. A traveler who has no connection with the characters in the novella narrates the story. This disconnect between narrator and story can make it difficult for the reader

E. I. Brooks et al. (Eds.): DLI 2020, LNICST 366, pp. 187–197, 2021.
https://doi.org/10.1007/978-3-030-78448-5_13

to engage with the story. The novella has therefore become notorious as one of the most boring and tedious texts in the mandatory Danish gymnasium curriculum.

This study's research question was as follows: Can a serious game be designed to increase reading engagement in St. St. Blicher's novella *The Hosier and His Daughter* as part of the mandatory reading in Danish literature classes at Ørestad Gymnasium?

Reading fiction is positively associated with higher performance on reading assessments [7]. However, in recent decades, students have changed their habits; they now read less fiction and spend more time reading online than before [2–4]. In Denmark, 20% of young adults do not read fiction [6], which is in line with other international reports [4, 7]. On average, across OECD countries, 37% of students report that they do not read for enjoyment at all [7]. Reading has always been encouraged through complex and diverse practices. However, there is considerable concern that young adults do not read well enough to cope with the increasing literacy demands of an information society [3–5]. Reading is a skill with many graduations of proficiency, and reading a rather complex novella from 1829 requires a different level of reading ability than, for example, reading a newspaper or subtitles on Netflix, making the former task challenging for many students [3–5]. If students read less fiction, they could lack familiarity with national literature of historical importance. In Danish gymnasiums, male students struggle more with reading engagement compared to female students [10]. This lack of reading engagement may partly explain why male students in particular lag behind compared female students in Danish gymnasiums, a fact that poses challenges for male students in terms of later educational opportunities and access to the labor market [8, 9]. In Danish gymnasiums, the average grade difference is 0.5 points (based on 7-point grading scale) in favor of female students [9]. However, in Danish literature, which includes a great deal of mandatory reading, female students' grades are 1.4 points higher on average [9]. Other countries are reporting similar phenomena such as higher dropout rates and lower average grades among male students [10]. Various sources have explained this in terms of genetic differences, different gender based learning identities, societal expectations and the feminization of the education sector [10]. A number of initiatives exist to increase reading engagement and decrease the gender gap in education, such as multiple initiatives related to serious games.

2   Previous Work and Theoretical Foundation

Scholars have described multiple principles for designing serious games for educational purposes [11–17, 26, 34], including a focus on reading engagement [10, 12, 13]. Important aspects of serious game design for learning and reading engagement include realism, feedback, discovery, repetition, guidance, flow, digital storytelling and debriefing [11–17]. Furthermore, motivation is important. Reading engagement, both in serious games and in other media, including analog media, requires the reader to be motivated [13, 16]. This involves aspects such as important elements within the text's content, text comprehension, knowledge acquisition and social interactions that employ knowledge and lessons learned from the text [13, 16]. Scholars have also emphasized the specific aspects of intrinsic motivation as important when designing serious games for reading engagement [16, 17]. These can include elements such as curiosity, a desire for a challenge, flow, involvement and narrative engagement [16–18]. In particular, narrative engagement [18] seems important within a serious game focused on reading engagement because of its relation to the story experienced while playing the game. Thus, it may result in imaginative immersion, narrative involvement or narrative immersion. The desire to know how the story of *The Hosier and His Daughter* unfolds evokes curiosity, suspense and narrative engagement, making the players want to continue playing [18].

Studies have also included transmedia storytelling as a gateway to reading engagement or educational purposes by combining analog reading with parts of the story included within a serious game [10, 19, 29–31]. The term transmedia storytelling is defined and used differently across disciplines. However, there is common agreement that it is a method or technique for telling a story across media platforms that often includes digital technologies such as serious gaming. Ryan, among others, stated that a successful digital transmedia storytelling requires a process of active collaboration in the co-creation of meaning [20].

3   Methods

3.1   Participants and Ethical Issues

This study in transmedia storytelling included 52 students from two Danish classes (M and S). Both classes were from Ørestad Gymnasium, which is located in Copenhagen, Denmark, and has a special profile focusing on media, communications and culture. Class M had a media studies profile and consisted of 27 students (14 male, 13 female). Class S had a social science profile and 25 students (12 male, 13 female). All participants gave informed consent and were informed that they could withdraw from the study at any time and their participation did not influence their grade. In addition, all participants were provided with anonymized ID numbers, and all data were labeled with these IDs. We applied special considerations when recruiting teenagers (ages 17–19), in accordance with Danish data law, the international code of conduct [23] and ethical approval from the gymnasium.

3.2   Procedure and Analysis

This study was conducted during the COVID-19 pandemic, so most of the study was conducted online. All teaching at Ørestad Gymnasium also took place online. The eight-step procedure used in this study is illustrated in Fig. 1. The procedure was divided into two major approaches. The pre-engagement and feedback stages (stages 1 and 8, and the hotline) focused mainly on the design and implementation elements. The reading and evaluation stages focused mainly on the evaluation elements.

Step 1: An important focus of this study was to involve the Ørestad Gymnasium teachers who taught the students about *The Hosier and His Daughter*. This was done by following a participatory design approach [22] in which the end-users included both teachers and students. The teachers served as gatekeepers who facilitated and controlled the reading process in areas such as the curriculum's aims, focus, knowledge, skills and analysis. Therefore, the teachers were involved as co-designers very early in the process.

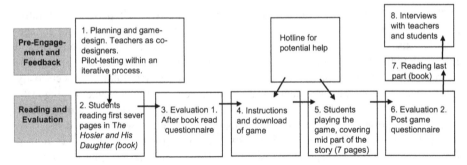

Fig. 1. An eight-step procedure used in this study.

They were asked for input and feedback, but they also worked as partners in the design process regarding changes to aspects of the game's development. Prior to the co-design efforts with the teachers, class observations and short interviews with students were conducted. Furthermore, the gymnasium's headmaster was interviewed during Step 1 and provided acceptance and engagement related to this study. Nine students from outside Classes M and S took part in pilot testing within an iterative process.

Step 2: Participants read the first seven pages of the novella *The Hosier and His Daughter* [1]. The novella was included in a textbook with other novellas and short stories. Prior to the reading, the teachers provided some minor instructions.

Step 3: The participants answered the first evaluation questionnaire after reading the novella's first seven pages. The questionnaire consisted of general questions related to the story and reading engagement at this stage. In total, 50 students completed Evaluation 1 (96% response rate).

Steps 4 and 5: The participants installed and played the game from a Google Drive folder. There were clear instructions for how to download the game. The game covered the part of *The Hosier and His Daughter* on pages 95–102 [1]. These specific pages were covered in the game due to an interesting time jump to five years later in the original story. This time jump in the narrative can challenge reader engagement, as the story at this point could be interpreted as having a natural end without any cliffhangers for a continuous read.

For Steps 4 and 5, a hotline was established so the students could contact the designers if they struggled with downloading the game, unforeseen bugs or other difficulties with the game. The hotline was used four times, and showed the importance to ensure that the students played the game and were not disengaged due to technical problems.

Step 6: The students answered the second evaluation questionnaire, which contained questions about their engagement after having played the game. The questionnaire had the following themes: learning, engagement and the game aspects; it was inspired by the User Engagement Scale [24]. For Evaluation 2, 20 participants from Class M (75% response rate), and 18 students from Class S participated (72% response rate). In total, 38 students from the two classes (16 males, 22 females), provided data for Evaluation 2.

Step 7: The students read the remaining pages of *The Hosier and His Daughter* (book).

Step 8: Interviews were conducted with two teachers and four students (three students from the M class, and one student from the S class). The semi-structured interviews had the following themes: esthetic appeal, endurability, involvement, focused attention, novelty and usability.

Researchers analyzed the questionnaires (Evaluations 1 and 2) using cumulative frequency (i.e., the total number of answers to specific questions). They analyzed the interviews using traditional coding [25] in four steps: organizing, recognizing, coding, and interpretation. They transcribed the interviews verbatim to be organized and prepared for data analysis. The researchers read the transcripts several times to recognize the concepts, which also included a general sense of the information and an opportunity to reflect on its overall meaning. Two researchers coded the interviews independently and then afterwards matched the themes derived, following the procedures to ensure inter-analysis reliability [32, 33]. The themes derived included engagement, learning, and game design. Researchers then categorized and interpreted each interview statement by following an interpretation of positive and negative statements within each of the three themes.

4 Design and Implementation

4.1 A Subsection Sample

The serious game was designed in Unity using C# for Windows, Mac and Linux. The game genre used within this study was an adventure puzzle solving game. To design the game, we followed eight flow principles described by Sweestser and Wyeth [21] to maintain the reading engagement. The eight elements of flow are as follows: concentration, challenge, skills, control, clear goals, feedback, immersion and social participation [21]. The students were able to start playing the game without reading a manual and learned the game controls in an introductory tutorial [21]. At the very start of the game, instructions appeared showing the keys to use for in-game navigation (Fig. 2, right). As the players discovered interactive objects, hints were shown, including which buttons to use for interaction. To promote concentration, this game implemented visual- and auditory stimuli using interactive objects that rewarded the player with a voice-over and written story text. The tasks (objects) needed to be completed in a specific order to ensure that the students read the story in order and followed the plot and original story. To highlight the reading objects, a particle system (Fig. 2, left) was implemented above the objects. The particles made it easier for the players to identify the objects needed to progress in the story. To avoid confusing players, the particles disappeared once activated. To evoke further engagement, sound effects were added when picking up clues or keys. This was done as players interacted with the clue to provide immediate feedback (Fig. 2, left).

Fig. 2. Instructions for navigation (right). Particle effects highlighting a reading object (left).

The students were expected to become less aware of their surroundings and more spatially immersed [17, 18, 21, 26] in the story, thereby improving their reading engagement for this difficult text. The immersive elements were implemented using visual in-game representations and settings that included assets appropriate for a house during the nineteenth century. In addition, the notes and letters the players had to find resembled old paper with text in a nineteenth-century-looking font. (Fig. 3, left). To further enhance the players' immersion, various ways of representing the text were implemented, though it appeared most often as text in an old book (Fig. 3, right).

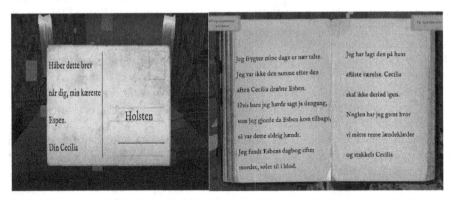

Fig. 3. The notes and letters the players had to find, which resembled old paper (left). Reading text presented in an old book format (right).

5   Findings

5.1   Reading Engagement and Learning

The findings from Evaluation 1 (after the analog seven-page reading) and from Evaluation 2 (after the reading within the serious game) revealed clear improvements in reading engagement within and after the gameplay compared to reading from the book (Table 1). Overall, 85% of the students found the in-game text easy or very easy to read compared to the book text, which only 16% considered easy to read (Table 1). In addition, 76% of the students enjoyed reading the story in the game to a large or very large extent, compared to 24% for the book. After playing the game, 55% of the students indicated that they would like to continue reading the story to a large or very large extent, compared to 36% after reading the book. Moreover, after playing the game, 56% of the students were immersed in the story to a large or very large extent, compared to only 14% after reading the book (Table 1).

Table 1. Results evaluation 1 (E1, after book read) and evaluation 2 (E2, after game play).

Results from Evaluation 1 (after book read, n = 50) and Evaluation 2 (after game play, n = 38). Numbers are in percentage	To a very small extent %	To a small extent %	To some extent %	To a large extent %	To a very large extent %
E1: I found the text in the novella easy to read	2	30	52	16	0
E2: I found the text in the game easy to read	0	5	10	27	58
E1: I enjoyed reading the story	8	32	36	24	0
E2: I enjoyed reading the story in the game	0	3	21	55	21
E1: I want to continue reading the story	8	28	28	32	4
E2: I want to continue reading the story	3	3	39	37	18
E1: I was immersed in the story	24	30	32	14	0
E2: I was immersed in the story	3	7	34	34	22
E2: The story affected me emotionally	22	24	36	10	8
E2: It was more enjoyable to read the story in the game than the book	3	5	8	34	50
E2: I wanted to know what happens to the characters in the game	0	3	18	55	24

The immersion question was phrased differently for Evaluations 1 and 2 to provide proper Danish language in its context. The immersion question addressed narrative immersion, so the same tendencies and validation of the students' immersion were revealed due their desire to know how the story would unfold. Overall, 79% of the students indicated that they wanted to know, to large or very large extent, what happens to the characters after playing the game. Narrative immersion may create curiosity, suspense and excitement, making the player want to continue reading [18]. Overall, 84% of the students perceived, to a large or very large extent that, that reading the story was more enjoyable in the game than in the book. This last question was not a direct indicator of reading engagement and involved potential bias (e.g., user, context and technology novelty effects). Nevertheless, it provided perceptions of enjoyment [21].

The game's ability to improve reading engagement and convey specific details surrounding the story's narrative was also a main focus of the teachers' feedback.

The potential with this serious game is there, along with the ability to motivate certain students to read longer texts that would otherwise be too difficult for them...a game can also go deep into the details in very specific parts of the text. (Teacher 1).

The game can make the students remember more from the story...and when you experience the story through a game, where you [the authors of this study] made the graphics, it controls how the students perceive the environment and the characters, just like when you watch a movie. When you don't have visual aids, you rely on your imagination, and you imagine the characters and the environment, so in one way or the other, you could say that it is a minor disadvantage (Teacher 2).

The quote from Teacher 2 emphasized the importance of being professional and providing the game environment with considerations, as it controls the perceptions. This is a well-known point from previous discussions of book-to-film adaptations. However, it is important to remember that serious games also work with messages and provide specific perceptions and interpretations based on their design and textual messages.

5.2   The Game Mechanics

The evaluations and interviews revealed that the students enjoyed the game. In addition, the game mechanics were effectively designed. Overall, 82% of the students always knew what to do in the game to a large or very large extent, and 79% found it easy to control the character to large or very large extent (Table 2). Furthermore, 78% of the students found it easy to follow the story. However, the interviews also revealed minor suggestions for improvements. In spite of the pilot testing and iterative adjustments to the game and tutorial, some students seemed to struggle with the controls and needed some time to get used to them.

Table 2. Results evaluation 2, game mechanics

Results from Evaluation 2 (after game play, n = 38). Numbers are in percentage	To a very small extent %	To a small extent %	To some extent %	To a large extent %	To a very large extent %
I always knew what to do in the game	0	0	18	42	40
I found it easy to control the character	3	0	18	37	42
I found it easy to follow the story	0	5	17	39	39
I found the game too easy	0	16	37	23	24
I was easily distracted while playing	16	47	26	8	3

As many other scholars have realized when designing serious games, it is especially challenging to find the right balance between skills and challenges, to keep the players in the flow state [28]. Some participants in this study would have liked the game to be a bit more challenging (Table 2), be less monotonous or tedious and feature better links, clues or semiotics to story-specific elements, as well as a customizable reading speed to provide dynamic difficulty adjustment.

6 Conclusion

To design a successful serious game for educational purposes, it is crucial to involve teachers in the design process. A serious game has the potential to bring a new level of understanding and visualization to a given story in the classroom. However, when focusing on literature (e.g., novels, novellas, poetry), books and games must supplement each other through the inherent advantages of their media formats. This is also emphasized in the literature [27], so it is beneficial to explore further possibilities related to transmedia storytelling. By switching between the text and the game, the students in this study experienced improved reading engagement, as well as a deeper understanding of the story's environments and characters. However, a game cannot provide insight into deeper subjects in the text or the vocabulary that the author uses, which are also significant aspects of the learning outcomes expected in Danish literature classes. In conclusion, this serious game, designed specifically to teach *The Hosier and His Daughter*, improved students' reading engagement, leading to much higher immersion and ease and enjoyment of reading. The story was well told in the game, and the learning outcome was achieved through increased engagement. However, future work is needed to create significant evidence and insights regarding students' reading engagement via transmedia storytelling. First, researchers need to include a much higher number of students from classes across different gymnasiums within the data collection. Second, they need further details on the identification of readers, including their confidence in reading. It is important to emphasize that there is no agreed taxonomy for reading engagement, and the inclusion of serious games is still diverse in its outcomes, and certainly understudied as a transmedia subject for inclusion in the Danish gymnasium. It would also be interesting to create different options in the game design to accommodate a diversity of students and reader types.

References

1. Blicher, St. St.: Hosekræmmeren [The Hosier and His Daughter][1829]. In: Møller, L., Thurah, T.: Kanon i Dansk 9. Gyldendal, Copenhagen (2005)
2. Baron, N.S.: Words Onscreen: The Fate of Reading in an Online World. Oxford University Press, Oxford (2015)
3. Ross, R.S., McKechnie, L., Rothbauer, P.M. (eds): Reading Still Matters: What the Research Reveals About Reading, Libraries, and Community. Libraries Unlimited, Santa Barbara (2018)
4. Twenge, J.M., Martin, G.N., Spitzberg, B.H.: Trends in US Adolescents' media use, 1976–2016: the rise of digital media, the decline of TV, and the (near) demise of print. Psychol. Pop. Media Cult. **8**(4), 329 (2019)

5. Cai, J., Gut, D.: Literacy and digital problem-solving skills in the 21st century: what PIAAC says about educators in the United States, Canada. Fin. Jpn. Teach. Educ. **31**(2), 177–208 (2020)
6. Book and literature panel Annual Report: Bogen og litteraturens vilkår 2018 [The book and litteratur 2018]. SLKS, Agency for Culture and Palaces. https://slks.dk/fileadmin/user_up-load/0_SLKS/Fotos/Bogpanel/Rapport18/Aarsrapport_2018.pdf. Accessed 21 July 2021
7. OECD 2010/PISA 2009: Results: Learning to Learn – Student Engagement, Strategies and Practices, vol. III. https://doi.org/10.1787/9789264083943-en
8. DEA: Gymnasiet taber drengene [boys lacks behind in the Gymnasium]. https://www.datocms-assets.com/22590/1586245825-deanotatgymnasiettaberdrengene.pdf. Accessed 21 July 2021
9. Statistics Denmark: Karaktergennemsnittet stiger for alle grupper af studenter (opdateret). https://www.dst.dk/da/ .Accessed 21 July 2021
10. Pasalic, A., Andersen, N.H., Carlsen, C.S., Karlsson, E.Å., Berthold, M., Bjørner, T.: How to increase boys' engagement in reading mandatory poems in the gymnasium: Homer's "The Odyssey" as transmedia storytelling with the Cyclopeia narrative as a computer game.. In: Guidi, B., Ricci, L., Calafate, C., Gaggi, O., Marquez-Barja, J. (eds.) GOODTECHS 2017. LNICSSITE, vol. 233, pp. 216–225. Springer, Cham (2018). https://doi.org/10.1007/978-3-319-76111-4_22
11. Zhonggen, Y.: A meta-analysis of use of serious games in education over a decade. Int. J. Comput. Games Technol. **2019**, 1–8 (2019)
12. Sylvester, A., et al.: Mapping learning and game mechanics for serious games analysis. Br. J. Edu. Technol. **46**(2), 391–411 (2015)
13. Naumann, J.: A model of online reading engagement: linking engagement, navigation, and performance in digital reading. Comput. Hum. Behav. **53**, 263–277 (2015)
14. Abdul Jabbar, A.I., Felicia, P.: Gameplay engagement and learning in game-based learning: a systematic review. Rev. Educ. Res. **85**(4), 740–779 (2015)
15. Vivitsou, M.: Digital storytelling in teaching and research. In: Tatnall, A., Multisilta, J. (eds.) Encyclopedia of Education and Information Technologies. Springer, Cham (2018). https://doi.org/10.1007/978-3-319-60013-0_58-1
16. Guthrie, J.T., Wigfield, A., You, W.: Instructional contexts for engagement and achievement in reading. In: Christenson, S., Reschly, A., Wylie, C. (eds.) Handbook of Research on Student Engagement, pp. 601–634. Springer, Boston (2012). https://doi.org/10.1007/978-1-4614-2018-7_29
17. Wouters, P., Van Nimwegen, C., Van Oostendorp, H., Van Der Spek, E.D.: A meta-analysis of the cognitive and motivational effects of serious games. J. Educ. Psychol. **105**, 249–265 (2013)
18. Schønau-Fog, H., Bjørner, T.: "Sure, I Would Like to Continue" a method for mapping the experience of engagement in video games. Bull. Sci. Technol. Soc. **32**(5), 405–412 (2012)
19. Kalogeras, S.: Media convergence's impact on education. In: Transmedia Storytelling and the New Era of Media Convergence in Higher Education, pp. 67–111. Palgrave Macmillan, London (2014). https://doi.org/10.1057/9781137388377_3
20. Ryan, M. (ed.): Narrative Across Media: The Languages of Storytelling. University of Nebraska Press, Nebraska (2004)
21. Sweetser, P., Wyeth, P.: GameFlow: a model for evaluating player enjoyment in games. Comput. Entertain. (CIE), **3**(3), 3–3 (2005)
22. Halskov, K., Hansen, N.B.: The diversity of participatory design research practice at PDC 2002–2012. Int. J. Hum. Comput. Stud. **74**, 81–92 (2015)
23. ICC/ESOMAR. https://www.esomar.org/what-we-do/code-guidelines.?page=what-we-do/code-guidelines. Accessed 21 July 2020

24. O'Brien, H.L., Cairns, P., Hall, M.: A practical approach to measuring user engagement with the refined user engagement scale (UES) and new UES short form. Int. J. Hum. Comput. Stud. **112**, 28–39 (2018)

25. Portaleoni, C.G., Marinova, S., ul-Haq, R., Marinov, M.: Data analysis and findings. In: Corporate Foresight and Strategic Decisions Lessons from a European Bank, pp. 130–246. Palgrave Macmillan , London (2013). https://doi.org/10.1057/9781137326973_7

26. Hafner, M., Jansz, J.: The players' experience of immersion in persuasive games: a study of my life as a refugee and PeaceMaker. Int. J. Serious Games **5**(4), 63–80 (2018)

27. Sande C., Michael, D.: Serious Games: Games That Educate, Train, and Inform. Muska & Lipman/Premier-Trade, Mason, USA (2015

28. Csikszentmihalyi, M.: Flow: The Psychology of Optimal Experience. Harper Perennial, New York (1990)

29. Raybourn, E.M.: A new paradigm for serious games: transmedia learning for more effective training and education. J. Comput. Sci. **5**(3), 471–481 (2014)

30. Gambarato, R., Dabagian, L.: Transmedia dynamics in education: the case of Robot Heart Stories. Educ. Media Int. **53**(4), 229–243 (2016)

31. Poy, R., García, M.: Wizards, elves and orcs going to high school: how role-playing video games can improve academic performance through visual learning techniques. Educ. Inf. **35**(3), 305–318 (2019)

32. Goodwin, L.D., Goodwin, W.L.: Are validity and reliability 'relevant' in qualitative evaluation research? Eval. Health Prof. **7**(4), 413–426 (1984)

33. MacPhail, C., Khoza, N., Abler, L., Ranganathan, M.: Process guidelines for establishing Intercoder reliability in qualitative studies. Qual. Res. **16**(2), 198–212 (2016)

34. Báldy, I.D., Hansen, N., Bjørner, T.: How to design and evaluate a serious game aiming at awareness of therapy skills associated with social anxiety disorder. In: Proceedings of the 6th EAI International Conference on Smart Objects and Technologies for Social Good, pp. 156–162 (2020)

Towards Applying ARCS Model for a Blended Teaching Methodologies: A Quantitative Research on Students' Motivation Amid the COVID-19

Usman Durrani[1]([⊠]) [iD] and Muhammad Mustafa Kamal[2] [iD]

[1] College of Engineering and IT, Ajman University, Ajman, UAE
[2] School of Strategy & Leadership, Coventry University, Coventry, UK

Abstract. A well-reputed course prepared according to sound instructional design principles and successfully delivered multiple times in a traditional face-to-face classroom mode failed to stimulate students' motivation to learn in an online delivery mode amidst the COVID-19 outbreak. Therefore, a motivational framework developed according to the processes outlined in the ARCS model, implemented, and tested using a single-case study. A cohort of seventy-five undergraduate students aged between 24 to 29 years from different program majors enrolled in a six-week mandatory IT in Business course participated in this research. A blend of a traditional flipped classroom and gamified teaching methodologies were applied in alignment with the ARCS model's four motivational factors: attention, relevance, confidence, and satisfaction, associated process, and strategies. Before, during, and after treatment surveys based on the original Instructional Material Motivation Survey (IMMS) with 36 questions were conducted to determine the effectiveness of blended teaching methodologies on students' motivation. As a result, the teaching resources of the selected course were systematically aligned as required. We found that the blended teaching methodologies based on the ARCS model, process, and strategies have enhanced and/or sustained students' motivation and kept the subject interesting in an online environment, and ultimately improved their learning.

Keywords: Gamification · Flipped classroom · Blended learning · ARCS model · COVID-19 · Online education · Digital education · Higher education

1 Introduction

We know the year 2020 for its disastrous COVID-19 outbreak and its global impact on every aspect of human life. In education, it has created significant challenges for the global education community. It is a test of organisational agility [21], where institutes initially focused on shifting their program delivery to online environments, not necessarily on online pedagogy [5]. Researchers have conducted studies to analyse the responses

E. I. Brooks et al. (Eds.): DLI 2020, LNICST 366, pp. 198–207, 2021.
https://doi.org/10.1007/978-3-030-78448-5_14

taken by different education providers globally, with solutions varying from no response through to an online offering [5, 17].

In the mid of March 2020, we had to suspend our regular face-to-face classes under the emergency policy by the UAE government. In these unfamiliar circumstances, there was uncertainty and disparity between what to teach, how to teach, where to teach, how to assess and grade, the workload of lecturers and students, the teaching environment, and the implications for education equity [22]. The University faced several challenges under the newly introduced online policies included: the weakness of the e-learning infrastructure, the lack of lecturers' experiences in conducting online classes, the complex environment at lecturers' and students' homes, and so forth.

Out of these challenges, one of the key academic challenges, our well-reputed courses prepared according to sound instructional design principles and successfully delivered multiple times in a traditional face-to-face classroom environment, failed to stimulate students' motivation to learn in an online delivery mode. A fundamental question, "How to motivate students in an online environment?" and "How to have a sustainable motivation throughout our courses?"

Exploration of such questions leads to considering holistic models such as the time-continuum model [20], and Keller's ARCS model [10]. For this study, we followed the approach of the latter. The ARCS model represents four categories of motivational factors; attention, relevance, confidence, and satisfaction. The model includes sets of motivational enhancement strategies and a motivational design process with application to any instructional design models [10]. The ARCS model with its basis in the macro-theory of motivation and performance [8, 9] and grounded from the expectancy-value theory [16] has been validated from various aspects in many studies [12, 19].

Past studies have applied the ARCS model in instructional design models by combining it with a flipped-classroom approach [1] to determine the effect on the achievements, motivation, and self-sufficiency of the students [4]. These studies have found a flipped classroom as a positive contributor towards enhancing the motivation and self-sufficiency of the students. In a flipped-classroom instructional design model, the lecturer provides resources to the students for reading before class and then uses class time to engage students in various learning activities such as discussions on lecture specific topics, and collaboration tasks among peers [3].

Moreover, Toussaint and Brown [18] in their study have shown how to increase students' motivation in their course by applying the ARCS model and concepts of gamification [14] to develop Serious Games for mathematics learning. Another study identified the lack of the use of innovative teaching materials such as building simulation performance tools as key reasons for students' demotivation as they felt they could not able to compete with other professionals [6]. In line with the latter, Burke [2] highlighted on using game mechanics and experience design to digitally engage and motivate people to achieve their goals.

For this study, we have used the ARCS model, process, and strategies, blending it with traditional, flipped classroom and gamified teaching methodologies (from here now referred to as blended teaching methods - BTM) to deliver an online course. This paper aims to present a method to enhance and/or sustain students' motivation by systematically aligning the teaching resources of a selected course based on the three-staged students' feedback analysis (before-course, during-course, and after-course).

The hypothesis for this study is:

H1: ARCS model and strategies positively affect an online course delivery through BTM to enhance and/or sustain students' motivation.

- **H1a:** ARCS model and strategies positively affect an online course delivery through traditional teaching method with a difference in the students' motivation
- **H1b:** ARCS model and strategies positively affect an online course delivery through gamified flipped classroom teaching method with a difference in the students' motivation
- **H1c:** ARCS model and strategies positively affect an online course delivery through BTM with a difference in the students' motivation

This paper begins with an overview of the method that formed the basis of this study in Sect. 2. Section 3 then describes the research findings and discussion. Finally, we provide the conclusion, examine the study's ramifications, and explores potential areas for future research in Sect. 4.

2 Methodology

The participants for this study were 75 undergraduate students aged between 24 and 29 from different educational programs. We divided these participants into two classes taught by the same IT lecturer. Out of these, 35 were female and 40 were male students. The 26 participants were from Accounting and Finance programs, 17 from IT/IS programs, 28 from management programs, and four from the marketing programs. These participants represented the overall diversity in the university, except that there were no participants from groups with special needs.

We chose a mandatory IT in Business course to cover participants from a broad range of educational programs. This was the convenience sampling with the intention for it to be representative. Like [7], the sample selected with a concern to uncover a range of students in various programs experience unconventional learning approaches. The course represented three credit hours with four two-hour lectures per week for six weeks. We divided the delivery of the course into two phases. In Phase 1 (first three weeks), we delivered two lectures each week through a traditional teaching method using the Blackboard Collaborate and Moodle. At the end of each of the lectures, dedicated questions and answers time was provided to the students. In Phase 2 (last three weeks), we used a gamified flipped classroom teaching method for the same participants using the same tools above, followed by in-classroom tasks and activities. Figure 1 presents a BTM along with the assessments.

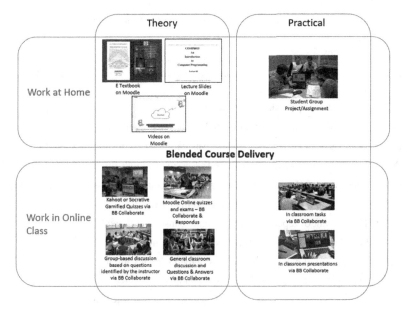

Fig. 1. Blended teaching methodologies (BTM)

The seven assessments performed along with their weightage during the BTM included: (1) one online formal midterm exam via the Moodle and the Respondus lockdown browser and monitoring tool – 20%, (2) one online formal quiz via the Moodle and the Blackboard Collaborate (Phase 1) – 10%, and (3) a team project and online presentation via the Moodle and the Blackboard Collaborate – 30%, (4) one gamified online quiz via the Kahoot or the Socrative and the Blackboard Collaborate (Phase 2) – 10%, (5) out-of-the-class activities per lecture (Phase 1), (6) in-class activities per lecture (Phase 2), and (7) one online formal final exam via the Moodle, the Respondus lockdown browser and monitoring tool – 30%.

A customised version of the Instructional Material Motivational Survey – IMMS (Table 1) measured students' motivation [9, 10], specifically learners' reactions to the motivational features of instructional material in terms of attention, relevance, confidence, satisfaction, and overall motivation. These questions were aligned with tools and techniques of flipped classroom and gamified teaching methodologies applied for this study. Internal consistency estimates for the IMMS total score and subscales are between 0.81 to 0.96 [15]. The IMMS used all 36 items from Keller's original 36 items, with twelve items for attention, nine for confidence, nine for relevance, and six for satisfaction. A five-point Likert-scale was used with a response choice of 1 (not true), 2 (slightly true), 3 (moderately true), 4 (mostly true), and 5 (very true). The responses were collected from the students at three stages of delivery: before-course, during-course, and after-course.

Table 1. IMMS survey 36 questions on ARCS motivational factors

Variable	Questions
ATTEN1	There was something interesting at the beginning of this course that got my attention
ATTEN2	Course delivery mode - is eye-catching
ATTEN3	The quality of the resources helped to hold my attention
ATTEN4	This course was so abstract that it was hard to keep my attention on it
ATTEN5	The design of formal and gamified assessments looks dry and unappealing
ATTEN6	The way the information is arranged in this course helped keep my attention
ATTEN7	This course has things that stimulated my curiosity
ATTEN8	The amount of repetition in this course caused me to get bored sometimes
ATTEN9	I learned some things that were surprising or unexpected
ATTEN10	The variety of classroom activities helped keep my attention
ATTEN11	The style of delivering lectures and conducting the assessments is boring
ATTEN12	There are so many words on each lecture slide that it is irritating
RELE1	It is clear to me how the content of this course is related to things I already know
RELE2	There were examples that showed me how this course could be important to people
RELE3	Completing flipped classroom activities during the sessions were important to me
RELE4	The content of this course is relevant to my interests
RELE5	There are explanations or examples of how people use the knowledge in this course
RELE6	The content and style of assessments used in this course convey the impression that its content is worth knowing
RELE7	This course was not relevant to my needs because I already knew most of it
RELE8	I could relate the content of this course to things I have seen, done or thought about in my own life
RELE9	The content of this course will be useful to me
CONF1	When I first looked at course, I had the impression that it would be easy for me
CONF2	The course was more difficult to understand than I would like for it to be
CONF3	After reading the introductory information, I felt confident that I knew what I was supposed to learn from this course
CONF4	Many of the slides had so much information that it was hard to pick out and remember the important points
CONF5	As I worked on this course, I was confident that I could learn the content
CONF6	The activities and assessments in this course were too difficult

(continued)

Table 1. (*continued*)

Variable	Questions
CONF7	After working on this course for a while, I was confident that I would be able to pass all course related assessments
CONF8	I could not really understand quite a bit of the material in this course
CONF9	The good organisation of the content helped me be confident that I would learn this material
SAT1	Completing the activities in this course gave me a satisfying feeling of accomplishment
SAT2	I enjoyed this course so much that I would like to know more about this topic
SAT3	I really enjoyed studying this course
SAT4	The feedback after the assessments, or of other comments in this course, helped me feel rewarded for my effort
SAT5	If felt good to successfully complete this course
SAT6	It was a pleasure to work on such a well-designed course

3 Findings and Discussion

For this study, we received 125 online survey responses from 75 students before-course, during-course, and after-course delivery. To decrease the chance of missing data, the Moodle learning management system was used to collect data from the enrolled students using a web-based questionnaire and set the required option for all the 36 questions (Table 1). The before-course responses were 35 out of 75 (total population), with a response rate of 47%. The during-course responses were 46 out of 75, with a response rate of 61%. For the after-course, 44 students responded with a response rate of 59%. The sum of average scores against each of the variables was then calculated and used to perform a test of normality and a paired t-test to their compared means.

To test the hypothesis that the results of before-course (M = 2.57, SD = 1.232), during-course (M = 3.07, SD = 0.364), and after-course (M = 3.07, SD = 0.289) were equal, we performed a paired t-test. Before conducting further analysis, we examined the assumption of normality distributed among different scores. The assumptions considered satisfied, as the skew and kurtosis level are estimated at -0.221 and -0.793, respectively, which is less than the maximum allowable values for a t-test (with skew < |2.0| and kurtosis < |9.0|) [13]. For H1a, the null hypothesis of equal means between before-course and during-course rejected, $t(35) = -2.314$, $p < 0.05$. Conversely, for H1b, the null hypothesis of equal means between during-course and after-course accepted, $t(35) = -0.154$, $p > 0.05$. The null hypothesis for H1c, with the equal means between before-course and after-course, was rejected, $t(35) = -2.381$, $P < 0.05$. Table 2 presents the adjusted 95% confidence intervals of the three pairs along with mean and standard deviations.

Table 2. Descriptive statistics of the three pairs in the study

	95% Confidence interval of the difference			Mean	Std. deviation
	Lower	Upper			
Pair 1	−0.942	−0.062	Before-course	2.57	1.232
			During-course	3.07	0.364
Pair 2	−0.062	0.053	During-course	3.07	0.364
			After-course	3.07	0.289
Pair 3	−0.937	−0.074	Before-course	2.57	1.232
			After-course	3.07	0.289

Before the course delivery, the students were very concerned and demotivated about the online mode of delivery and assessments. This was because of unfamiliarity with the e-learning education paradigm, weakness of the e-learning infrastructure, the complex environment at lecturers' and students' homes. Based on the before-course students' evaluation, in Phase 1, the lecturer selected customised strategies aligned with the ARCS model such as Blackboard Collaborate polling for student attention, taking examples from students' areas during lecture delivery to promote relevance, clearly identifying the rubrics for online assessment and publishing it on Moodle to boost students' confidence, and real-time projects based on students' choices with guidelines for the satisfaction. Because of these strategies, students' motivation shown visible improvement in terms of motivation (pair 1).

At the end of Phase 1, the same students went through during-course evaluation to assess students' motivation. There was a significant improvement in the students' motivation in terms of attention, relevance, confidence, and satisfaction. As a result, in Phase 2, the lecturer has used a distinct set of customised strategies aligned with the ARCS model to sustain students' motivation. This included using gamification tools for in-class MCQ assessments using the Kahoot and/or the Socrative tools for attention, scenario-based tasks during flipped classroom group breakout sessions to encourage relevance, multiple gamified assessments, and the selection of the best attempts towards final grading to boost confidence, and immediate feedback during the flipped classroom session for formal assessments. This sustained students' motivation is clear from pair 2 in Table 2.

At the end of the course delivery, the lecturer evaluated the students one last time to assess if the blended delivery (Phase 1 + Phase 2) of the course has positively affected the overall students' motivation. The overall effect of BTM was positive on the students' motivation (pair 3). Figure 2 illustrates the four categories of ARCS motivational factors and systematic alignment based on the analysis of the students' motivation before-course, during-course, and after course delivery.

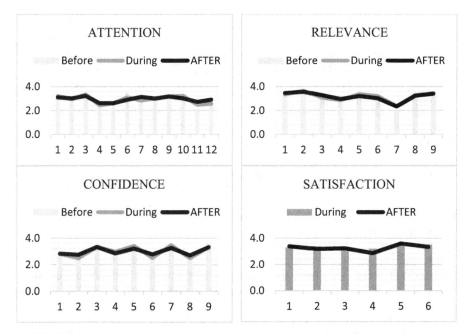

Fig. 2. Measurement of ARCS motivational factors - before, during and after the study

4 Conclusion, Limitations and Future Work

As a lecturer or designer, if motivational challenges or deficiencies arise, as it did in our study because of the COVID-19 outbreak, then it is necessary to first plan a motivational approach that will overcome these problems [11]. Students' motivation requires enhancement only when they are de-motivated [15]. However, if students' overall motivation is already high, then the lecturer or designer only requires sustaining students' motivation by using variety in teaching approaches [11] rather than exposing them to any unnecessary enhancement motivational tactics [9, 10]. In this study, we have shown the blended use of traditional, flipped classroom and gamified teaching methodologies in alignment with the ARCS model, process, and strategies with an adaptive implementation based on students' feedback before-course, during-course, and after-course delivery.

Based on the analysis of 125 responses collected from 75 students during three stages of the course delivery (before-course, during-course, and after-course), we conclude that an adaptive implementation of BTM aligned with ARCS models, process, and associated strategies can enable the enhancement and/or sustainability of students' motivation. The lecturers or designer should identify the right combination of BTM and activities through iterative motivational feedback of the students during course delivery.

The limitation of this paper is its convenient sampling through enrolled undergraduate students of the Business in IT course. It does not consider different courses, programs, colleges, institutes, and educational levels. As a result, the authors cannot generalise the findings of this paper to other research contexts.

Future work will focus on applying the recommendations presented in this paper to enhance courses in different programs in other colleges and educational levels. Besides, we are also planning to apply virtual reality technology in a project management course by applying the same BTM as followed in this paper. This will further support investigating the effects of BTM on the motivation of students to developing their skills in areas other than IT in Business.

References

1. Baker, J.: The 'classroom flip': using web course management tools to become the guide by the side. In: 11th International Conference on College Teaching and Learning, Jacksonville, FL (2000)
2. Burke, B.: Gamify: How Gamification Motivates People To Do Extraordinary Things, Bibliomotion, Brookline, MA (2014)
3. Butt, A.: Student views on the use of a flipped classroom approach: evidence from Australia. Bus. Educ. Accred. 6(1), 33–43 (2014)
4. Chang, Y.H., Song, A.C., Fang, R.J.: Integrating ARCS model of motivation and PBL in flipped classroom: a case study on a programming language. Eurasia J. Math. Sci. Technol. Educ. 14(12), em1631 (2018). https://doi.org/10.29333/ejmste/97187
5. Crawford, J., Butler-Henderson, K., Rudolph, J., Glowatz, M.: COVID-19: 20 countries' higher education intra-period digital pedagogy responses. J. Appl. Teach. Learn. 3(1) (2020). https://doi.org/10.37074/jalt.2020.3.1.7
6. Fernandez-Antolin, M.M., del Río, J.M., Gonzalez-Lezcano, R.A.: The use of gamification in higher technical education: perception of university students on innovative teaching materials. Int. J. Technol. Des. Educ. 1–20 (2020). https://doi.org/10.1007/s10798-020-09583-0
7. Goodchild, T., Speed, E.: Technology enhanced learning as transformative innovation: a note on the enduring myth of TEL. Teach. High. Educ. 24(8), 948–963 (2019)
8. Keller, J.M.: Motivation and instructional design: a theoretical perspective. J. Instr. Dev. 2(4), 26 (1979). https://doi.org/10.1007/BF02904345
9. Keller, J.M.: Motivational design of instruction. In: Reigeluth, C.M. (ed.) Instructional Theories and Models: An Overview of Their Current Status, Lawrence Erlbaum Associates, New York (1983)
10. Keller, J.M.: Development and use of the ARCS model of instructional design. J. Instr. Dev. 1(3), 2–10 (1987)
11. Keller, J.M.: How to integrate learner motivation planning into lesson planning: the ARCS model approach. In: VII Semanario, pp. 1–13. Santiago, Cuba (2000)
12. Loorbach, N., Peters, O., Karreman, J., Steehouder, M.: Validation of the instructional materials motivation survey (IMMS) in a self-directed instructional setting aimed at working with technology. Br. J. Edu. Technol. 46(1), 204–218 (2015). https://doi.org/10.1111/bjet.12138
13. Posten, H.O.: Robustness of the two-sample T-test. In: Rasch, D., Tiku, M.L. (eds.) Robustness of Statistical Methods and Nonparametric Statistics. Theory and Decision Library (Series B: Mathematical and Statistical Methods), vol. 1. Springer, Dordrecht (1984). https://doi.org/10.1007/978-94-009-6528-7_23
14. Prensky, M.: Digital Game-Based Learning. McGraw-Hill, New York (2001)
15. Song, S.H., Keller, J.M.: Effectiveness of motivationally adaptive computer-assisted instruction on the dynamic aspects of motivation. Educ. Tech. Res. Dev. 49(2), 5 (2001)
16. Tolman, E.C.: Purposive Behavior in Animals and Men, University of California Press, Berkeley (1932)

17. Toquero, C.M.: Challenges and opportunities for higher education amid the COVID19 pandemic: the Philippine context. Pedagog. Res. **5**(4), em0063 (2020)
18. Toussaint, M.J., Brown, V.: Connecting the arcs motivational model to game design for mathematics learning. Transformations **4**(1), 19–28 (2018). https://nsuworks.nova.edu/transform ations/vol4/iss1/3
19. Visser, J., Keller, J.M.: The clinical use of motivational messages: an inquiry into the validity of the ARCS model of motivational design. Instr. Sci. **19**, 467–500 (1990)
20. Wlodkowski, R.J.: Enhancing Adult Motivation to Learn: A Comprehensive Guide for Teaching All Adults, Rev Jossey-Basss, San Francisco (1999)
21. Wu, Z.: How a top Chinese University is responding (2020). Retrieved from World Economic Forum https://www.weforum.org/agenda/2020/03/coronavirus-china-the-challenges-of-online-learning-for-universities/
22. Zhang, W., Wang, Y., Yang, L., Wang, C.: Suspending classes without stopping learning: China's education emergency management policy in the COVID-19 outbreak. J. Risk Finan. Manag. **13**(55) (2020). https://doi.org/10.3390/jrfm13030055

Author Index

Printed in the United States
by Baker & Taylor Publisher Services